W9-CMM-099

Troubleshooting and Repairing Heat Pumps

Other McGraw-Hill books by the author

Building Contractor: Start and Run a Money-Making Business
Home Plumbing Illustrated
HVAC Mechanic: Start and Run a Money-Making Business
Journeyman Plumber's Licensing Exam Guide
Master Plumber's Licensing Exam Guide
The National Plumbing Codes Handbook
The Plumber's Troubleshooting Guide
The Plumbing Apprentice Handbook
Plumbing Contractor: Start and Run a Money-Making Business
Roofing Contractor: Start and Run a Money-Making Business
Security Systems Technician: Start and Run a Money-Making Business
Siding Contractor: Start and Run a Money-Making Business

Troubleshooting and Repairing Heat Pumps

R. Dodge Woodson

TAB Books

Division of McGraw-Hill, Inc.

New York   San Francisco   Washington, D.C.   Auckland   Bogotá
Caracas   Lisbon   London   Madrid   Mexico City   Milan
Montreal   New Delhi   San Juan   Singapore
Sydney   Tokyo   Toronto

pbk   1 2 3 4 5 6 7 8 9 0 DOC/DOC 9 9 8 7 6 5 4
hc   1 2 3 4 5 6 7 8 9 0 DOC/DOC 9 9 8 7 6 5 4

Library of Congress Cataloging-in-Publication Data

Woodson, R. Dodge (Roger Dodge), 1955–
   Troubleshooting and repairing heat pumps / by R. Dodge Woodson.
   p.   cm.
   Includes index.
   ISBN 0-07-008612-5 (H)   ISBN 0-07-008613-3 (pbk.)
   1. Heat pumps—Design and construction. I. Title.
TJ262.W66 1994
621.402'5—dc20   94-29455
                                    CIP

Acquisitions editor: April D. Nolan
Editorial team: Andrew Yoder, Managing Editor
               Melanie D. Holscher, Editor
Production team: Katherine G. Brown, Director
                Susan E. Hansford, Coding
                Rose McFarland, Desktop Operator
                Lorie L. White, Proofreading
                Joann Woy, Indexer   GEN1
Designer: Jaclyn J. Boone, Designer   0086133

To Kimberley and Afton, the people who mean the most to me.

Acknowledgments

I thank Dawn Gearld-Hall for her administrative support in putting this project together. Dawn's efforts have allowed me to put these pages into your hands faster.

I would also like to thank my parents, Maralou and Woody, for their many years of help in all of my endeavors.

The following companies were gracious enough to supply illustrations for this book, which make it easier for you to understand how to work with heat pumps:

Bard Manufacturing Company
Crispaire Corporation
Friedrich Air Conditioning Company
The Trane Company, an American-Standard company

Contents

Introduction

If you're thinking of buying a new heat pump, there are many things to consider. Will you purchase an air-source heat pump or a water-source unit? Can you save money by doing the work yourself? Is a heat pump really as efficient and cost-effective as you've heard? These are just a few questions that will come up during your decision-making process. Fortunately, you are holding the answer to all of these questions, and a lot more. This book is packed with information that anyone who is planning to work with a heat pump should have.

I've written this book in a friendly, conversational tone, refraining from the dry, hard-to-understand manner that you would expect to find in a training manual. This book is very much a training manual, but it's a lot easier to read and comprehend than most how-to books.

What can you learn from this book? After reading the chapters that follow, you will have a wealth of knowledge on the subject of heat pumps. You will understand how they work, why they work, and what can be done when they don't work. Having this book by your side as you work with your heat pump will make the job easier and less frustrating.

Take a few moments to look over the table of contents. Then thumb through the pages and look at the many illustrations that compliment the text. Sometimes a picture is worth a thousand words, and there are dozens of illustrations sprinkled throughout these pages.

1

Is a heat pump right for you?

Almost everyone has heard of heat pumps, and many people swear by them. Others swear at them. Who's right?

Heat pumps have been in existence for decades. Still a large number of people doubt the efficiency so often raved about when describing the operation of heat pumps. Are these people just die-hard skeptics, or are they correct in their lack of enthusiasm?

Heat pumps can be extremely efficient, both as a heating unit and as an air-conditioning unit. The cost of operating a heat pump can be considerably less than the expense of producing equivalent heat with electric heat or oil heat. Given the many types, models, and configurations of heat pumps available, a heating and cooling system can be designed to give maximum comfort in climate control and cost-effectiveness.

The key to having a favorable experience with a heat pump is knowledge. A heat-pump system that works wonderfully in North Carolina might not fare so well in Maine. It is not practical to assume one type of heat-pump system will work in all parts of the country. There are, however, enough options available when designing a heat-pump system to build a suitable system for almost any need.

If you're thinking of installing a heat pump, you have a lot to consider. Not taking the time to research your options will most likely result in disappointment. The subject of heat pumps does have many facets, but these many angles are to your advantage. Having so many ways to build a heating-and-cooling system allows you to customize a system to meet your needs in the most economical and efficient manner.

Working through the choices available with heat pumps is not a complex problem if you have help and guidance from a knowledgeable source, and this book is that source. This guide walks you step by step through the subject of heat pumps.

I've worked in the construction trades for twenty years. My career started as a plumber and evolved into much more. My first business was a plumbing and heating business. That business grew to include full-scale remodeling jobs, and from there, the business went on to include building up to 60 single-family homes a year, most of them equipped with heat pumps.

As a plumbing and heating contractor, I've installed and worked on many types of heating equipment. Hot-water baseboard heat, radiators, forced-hot-air systems, wall-mount heaters, and heat pumps have all been a part of my working environment.

With the experience and knowledge that I have gained in the construction field, why would I equip so many of the houses I've built, including homes for my personal use, with heat pumps? The reason is simple, heat pumps, in my opinion, are one of the most effective means of providing heating and cooling services for a home.

As much as I like heat pumps, I must admit that they are not always the best choice when debating a new heating or cooling system. States with extremely cold climates, like Maine, are not ideal locations for heat pumps to do their best work. This is not to say that heat pumps can't be effective in extremely cold temperatures, but there are several factors to consider in making the right choice in such a climate. I don't want to get too technical yet, this chapter is meant to give you an overview of heat pumps, but let me give you an example to consider.

Assume that it is a winter day. The temperature in Virginia is 20 degrees F. The temperature in Maine is also 20 degrees F. If identical houses, both with identical heat pumps, existed in Maine and Virginia, would both of the heat pumps be equally effective on this winter day? They might, but there is a good chance that the heat pump in Maine would be forced to work much harder than the heat pump in Virginia on this particular day. How could this be when the temperature in both locations is the same?

Maine typically receives much more snow than Virginia does, and falling snow and moisture in the air can make a big difference in the performance of a heat pump. If it were snowing in Maine, the heat pump might have to defrost its coil every hour. The same heat pump in Virginia, where it is not snowing, might only have to defrost its coil once a week.

As we progress into future chapters, I'll give you more of these examples and complete, technical details for working with heat pumps.

New construction

New construction of a home is an ideal opportunity for taking advantage of the benefits a heat pump can offer. If you are planning to build a new home, the options for using a heat pump are numerous.

When building a new home, you can install insulation easily. Having energy-efficient windows and doors installed is not difficult, and you can plan provisions for the duct work needed with a heat-pump system.

Before your new home is built, a heat-gain and heat-loss schedule can be worked up to show exactly what your heating and cooling needs are expected to be. With the results of such a test, you can modify your construction plans to maximize the efficiency of your heat pump. New construction certainly provides ample opportunity to make the most of a heat pump.

Remodeling

Remodeling a home does not offer the same freedom for capitalizing on the use of a heat pump that building a new home does, but extensive remodeling might provide the chance to install a heat pump. Many old heating systems work at an efficiency rating of only 60 to 70 percent. The desire for a more cost-effective heating-and-cooling system often points remodelers in the direction of heat pumps.

Old houses are frequently poorly insulated and have drafty windows and doors. The glass in old windows is not of the same insulating quality that is available in new windows. A number of older homes are not equipped with duct work, and the ones that are often have small ducts.

Heat pumps require fairly large ducts to be effective. A house with undersized duct work, or no duct work at all, can make the installation of a heat pump a sizable job. Air infiltration around windows and doors hurt the performance of any heating system, and inadequate insulation doesn't allow for maximum heating and cooling comfort at affordable prices.

Simply installing a new heat pump in place of your old furnace could result in frustration and dissatisfaction. If you are planning to replace an old system with a new heat pump, or even if you are planning only to add a heat pump to your existing system, there will be many thoughts to consider first. These considerations are discussed later.

Terminology

It is always helpful to understand the jargon used in the trade when you are dealing with contractors and suppliers. Here are a few definitions:

One-piece units

One-piece units are heat pumps that are self contained. You've probably seen this type of unit in motels. To make it easy to visualize, think of a one-piece heat pump as a window air conditioner that is also capable of blowing heat.

One-piece units are normally installed in a roof or a wall. This type of heat pump is typically used to heat and cool specific areas, rather than whole houses. If, for example, you were adding a family room to your home, a one-piece heat pump might make a viable choice for heating and cooling the space.

Two-piece systems

Two-piece systems are comprised of two primary pieces of equipment. One piece of the heat pump is installed inside the home, usually in a closet or similar space. The other piece of the heat pump is installed outside of the home, usually on a concrete pad that sits on the ground. The two pieces are connected with pipes that transfer the heat. This is the type of heat pump most often used in residential applications.

Heat sources

The heat sources for heat pumps are varied and numerous. Here's a look at each source of heat you might want to consider:

Air Air is a good source of heat, and it is the most common source used for heat pumps in the United States. In states that maintain moderate winter climates, air-source heat pumps work fine. If, however, the outside temperature dips into single digits, these systems lose their effectiveness, and a back-up heat source (usually electric heat strips) are required.

Well water Well water can be an excellent heat source. I know it might seem strange to think of getting your home's heat from a water well, but well water is a very good source for heat pumps to work with.

Surface water Surface water can be used as a heat source, but it ranks low on the list of possible choices. Surface water, and all the other heat sources, are discussed in greater detail later.

Earth The earth is a good source of heat for heat pumps, but there are some disadvantages to earth systems, too. As you will find later in the book, there is no single choice in heat pumps and heat sources that will serve all needs.

Solar Solar heat sources are also good for heat pumps. The costs associated with installing for solar heat can be extreme, and there are other considerations to be examined before a solar source is chosen.

The advantages

Let's look now at some of the advantages of a heat pump. Assuming that the proper research has gone into the design and installation of a heat pump, the system should be less expensive to operate than other types of heating systems. In mild to moderate climates, heat pumps are at their best. Extreme climates push heat pumps to their extremes, and the units are not always up to the challenge.

Sources of heat for heat pumps to pull from are numerous. Because heat pumps can be configured to use any of the various heat sources, they can be used under varied conditions.

Safety is another advantage gained when heating and cooling with a heat pump. There is no fuel oil to burn, no fumes that must be vented from the home, and no risk of a heating fire escaping its chamber.

A big advantage of a heat pump is its ability to double as both a heating and a cooling unit. There is no need for a furnace and an air-conditioner when you install a heat pump—the heat pump is both.

The space-saving quality of a heat pump is an advantage over some types of heating systems. Because the only portion of the heat pump that takes up space in your home can be concealed in a small closet, it requires much less space than heating systems that demand the storage of an oil tank and considerable floor space for the furnace.

The disadvantages

If you buy a quality system that has been properly designed for your home, there is very little to consider that might be termed a disadvantage.

Because a heat pump becomes less effective as the temperature of its heat source drops, you might make a case for this being a disadvantage. Most heat pumps pull their heat source from air. When the outside air temperature drops, homeowners want the interiors of their homes to remain warm. The heat pump loses some of its heat producing ability as the temperature gets colder, which is when the heating aspect of the heat pump is needed most, resulting in a possible disadvantage. However, if the heat-pump system has been designed and installed properly, this "disadvantage" should not come into play.

People in colder climates might find the initial cost of a heat-pump system to be a disadvantage. It is not unusual for a heat-pump installation to cost more than a standard furnace, but remember, you are getting not only a furnace, but a central air-conditioning system as well. Because many homes in colder climates, such as Maine, don't have or need central air conditioning, the additional cost for a heat pump might not be warranted.

Is a heat pump right for you?

There is still much to consider before you should begin to try making that decision. The odds are good, however, that a heat pump is right for you.

I have lived in homes with electric heat, wood heat, hot-water baseboard heat, forced hot-air heat, radiators, and heat pumps. Allow me to share my feelings about each of these types of heat. My personal experiences might help you in making your decision about heat pumps.

Electric heat

The homes I've lived in that were equipped with electric heat, both in Virginia and Maine, have been very expensive to keep warm. Electric heat rarely feels warm, even when the thermostat indicated an interior temperature that should have been more than comfortable. Electric heat produces heat quickly, but it is expensive to operate.

Wood heat

I love wood heat. The aroma of wood smoke, the warmth, and the sense of going back in time all combine to make wood heat a favorite of mine. However, as I've gotten older, carrying wood in for the stove seems to get more difficult with each trip to the barn. Not only is carrying wood physically demanding, the wood leaves debris throughout the house. It is impossible to clean out the stove without creating a dust layer, and with a small child in the home, a wood stove can be a disaster waiting to happen. There is also the danger of unwanted fire and the expense of having the chimney cleaned.

In my younger days, I heated my home exclusively with wood. During those years I didn't mind the mess and the trouble. Now, I use wood only for the sense of pleasure it gives me to see it burning in the fireplace. The warmth is still welcomed, but the wood stove is gone and I don't depend on wood to heat my home any longer.

Hot-water baseboard heat

Hot-water baseboard heat is the heat of choice in Maine. More homes have this type of heating system than any other. Hot-water baseboard is good for Maine's extreme climate, but the boilers and oil tanks take up considerable space, there are oil deliveries that must be made on snowy roads, and the heating system does nothing to cool the home during those few weeks of the year when it is hot in Maine.

Forced hot-air heat

Forced hot-air heat is common. Some people don't like it because of the dust it moves around, but in general, it is a well-accepted form of heat. My experience with forced hot-air heat in Maine has been acceptable, but somewhat expensive. When you consider that a heat pump can give heat and air conditioning through the same ducts, it makes little sense to use only a forced hot-air furnace when a heat pump could be used.

Heat pumps

All of the new homes I've built for myself have been equipped with heat pumps. The benefit of having my heating and air-conditioning needs met by a single piece of equipment has always made me happy. The performance of the heat pumps in my homes have always been satisfactory. Maintenance on the heat pumps have been low, and the units have never forced me to live in a cold house.

Of all the heating systems available, I would opt first for a heat pump and second for a hot-water baseboard system. This choice is based on overall average conditions. In extreme conditions, I might lean more towards the hot-water baseboard for one reason: the resale value of the home. In some areas, such as Maine, home buyers have preconceived opinions of value, and many rank hot-water baseboard heat higher than heat pumps. As far as overall cost, efficiency, comfort, and use, I prefer heat pumps.

As you might recall, I told you earlier that I have built as many as 60 single-family homes in a single year. Most of my building activity was carried out in Virginia. The winter temperatures in the part of Virginia where I worked frequently went into the teens and sometimes much lower. Summer temperatures hanging in the upper 80s and 90s were common.

A state where the cold temperature can hit zero and the hot temperature can exceed 100 degrees F. should be a good testing ground for heat pumps. I can tell you from experience, not one of my home-buying customers ever complained about the performance of their heat pumps in summer or in winter. I think this says a lot for heat pumps.

In Maine, there are extremely cold days during the winter. The summers are typically cool, so air conditioning in homes is not a requirement. These two factors combine to make heat pumps less prevalent in Maine. However, commercial buildings in Maine, where air-conditioning is a needed convenience, do use heat pumps. Do you think savvy investors and real estate developers would invest big bucks in systems that can't handle extreme cold? I don't, and I know, again from first-hand experience, that heat pumps do work in Maine.

All in all, if you are seeking an efficient way to heat and cool your home, a heat pump is probably your best choice. The next chapter explores the mysteries, myths, and misconceptions on the effectiveness of heat pumps in various parts of the country. Following chapter 2, is a journey into the technical world of heat pumps. By the time you finish this book, you will have all the information you need to make an informed decision on whether or not a heat pump is right for you. You might find the symbols and abbreviations in Table 1-1 helpful during your work with heat pumps.

Table 1-1 Symbols and abbreviations that might be encountered

Btu	British Thermal unit
Btuh	BTU per hour
CMF	Cubic feet per minute
D.D.	Degree day
E	Efficiency
ESP	External static pressure
GPM	Gallons per minute
H	Convective heat transfer coefficient
HL	Heating load
IN WG	Inches of water gauge pressure
KW	Kilowatt
KWH	Kilowatt hour
MCF	1,000 cubic feet

OAT	Outside air temperature
Q	Heat flow
R	Thermal resistance
RAT	Room air temperature
T.D.	Temperature difference
TON	12,000 Btu per hour
V	Volts

2

When is a heat pump cost effective?

Most of the time, heat pumps are cost effective. It is actually easier to talk about when a heat pump is not cost effective.

The efficiency of heat pumps is effected by many factors. The one most often associated with the effectiveness of a heat pump is the climate. People assume that if a climate becomes frigid in the winter, heat pumps will not be cost effective. Sometimes this is true, but there are many other elements that contribute to the cost effectiveness of a heat pump. Some of the factors are of a technical nature, and some aren't.

This chapter examines facts and figures about heat pumps. It also shows you some aspects of assessing cost effectiveness that you might not expect to find. The technical side of the question is addressed, along with the human side of it. This chapter is intended to give you a broad understanding of what makes a heat pump cost effective. It also explains the circumstances that can defeat the effectiveness of a heat pump, and what you might be able to do in overcoming those problems.

A simple question

A simple question that many people tend to overlook when thinking about the cost effectiveness of a heat pump is one of whether or not they need or desire central air conditioning. If you have no need for air conditioning, the acquisition and installation cost of a heat pump is not going to make much sense.

If you compare the cost of a heat-pump installation to a standard furnace, you will likely find the heat pump to cost somewhat more. The increased initial cost might discourage you from purchasing a heat pump. This is a logical assumption, but you should also consider the cost of operating the two different types of heating equipment.

If you plan to live in your home for a number of years, there is a good chance that the heat pump will save enough in operation costs to make its more expensive purchase price worthwhile, and you will have the added benefit of air conditioning when you want it. Too many people see only the larger initial cost and fail to consider the overall cost in the long run.

Operating costs

There are two separate cost considerations when estimating the expense of operating a heat pump. One is the cost of heating your home, and the other is the cost of cooling it. If you can project what the operating costs for different types of heating systems will be in your area, you can determine, accurately, the cost effectiveness of a heat pump.

Coefficient of performance (COP)

The coefficient of performance (COP) is an industry standard for rating the effectiveness of a given heat pump. Basically, the higher the COP number is, the more efficient the heat pump is. Just as an insulation with an R-value of 19 has better insulating quality than an insulation with an R-value of three, a heat pump with a COP of three is more efficient that a heat pump with a COP of 2.5. Knowing this little bit of information makes comparing heat pumps to other types of heating equipment, and even to other heat pumps, much easier.

You can put the COP rating of a heat pump into direct comparison with other types of heating equipment. There is one thing to keep in mind when doing your comparisons. The COP rating of a heat pump is usually established with test results being based on an outside temperature of 47 degrees F and an inside temperature of 70 degrees F.

To put COP ratings to use, you must have a good idea of what other types of heat cost. Electric heat, for example, typically gives a dollar's worth of heat for a dollar's worth of electricity. While a one-to-one ratio is not all bad, it's not as good as heat pumps in moderate climates.

Fuel oil is worse than electricity in its return on your investment. Because heating systems that burn oil do not operate at 100% efficiency, you have to do a little guess work on how much heat you are getting for your oil dollar. Average oil burners run with efficiency ratings that range from 60% and 80%. If you assume the best of an oil burner, your one dollar's worth of oil is giving you about 80 cents worth of heat.

If you buy a heat pump with a COP of 2.5, your dollar's worth of operating cost is giving you two dollars and fifty cents worth of heat. A heat pump with a COP of three will give you three dollars worth of heat for each dollar you spend in operating costs. With this type of performance, it is easy to see how heat pumps can pay for themselves over a period of time.

Electric heat compared to a heat pump

Electricity gives a one-on-one return on your heat investment. For the sake of comparison, let's set an annual heating expense of $1,200 when heating with electric heat. How much would you have to spend in operating cost to get $1,200 worth of heat from a heat pump with a COP of 3. The answer is simple; it would cost $400 to get the same amount of heat from the heat pump that you are paying $1,200 for with your electric heat.

This figure was derived by dividing the annual cost of electric heat ($1,200) by the COP rating of the heat pump (3). If the COP rating had been listed as a two, the cost of operating the heat pump would have been $600.

A heat pump with a COP of three, as described above, would save you $800 a year in operating costs. In five years, the savings would be $4,000. Ten years of use would result in a savings of $8,000. Using this type of research and planning, you can clearly define solid projections for the cost effectiveness of your heat pump.

Energy efficiency ratio (EER)

When the cooling effectiveness of a heat pump is measured, it is done with a rating of an energy efficiency ratio (EER). Just as is the case with a COP rating, the higher the number for the EER rating is, the better the heat pump is. Knowing how to read and understand COP and EER ratings will greatly increase your ability to judge heat pumps fairly.

Geographical considerations

Geographical considerations cannot be ignored when projecting the cost effectiveness of a heat pump. Some heat pumps can function properly at temperatures well below zero, but not all heat pumps have the same ability to deal with extremely cold temperatures. Some models carry manufacturer's recommendations that suggest the heat pump not be used if the outside temperature drops into the teens. Other makes are rated to go all the way down to 20 degrees below zero. If you live in a cold climate, you will want to investigate the recommended cut-off point for the heat pump.

There is a formula used by manufacturers to establish the seasonal performance factor (SPF) of a heat pump. The SPF can help you identify how efficient and cost effective a particular heat pump will be in a given part of the country. If you were looking at a heat pump with a SPF rating of a 2, the heat pump would provide twice as much heat for your money as electric heat would.

The SPF rating of a heat pump is tied tightly to seasonal temperatures. As an example, southern Maine has a SPF rating of 1.8 while Arkansas has a SPF of 2.25. Florida's SPF is 2.5. These numbers allow you to determine the cost effectiveness of the heat pump. Even in the coldest continental United States, the SPF runs 1.6.

Let's put this knowledge into the form of a comparison. Where I live in Maine, the SPF is 1.8. If we say my house would cost $2,000 a year to heat with electric heat, how much would it cost with a heat pump? The cost would be just over $1,100, a savings of about $900 a year.

Energy-savings charts on heat pumps show that they are expected to save anywhere from about 20% to about 60% on the heating of an average house. These figures are based on air-source heat pumps (the most common type) and resistance heating.

The fluctuation in the savings is linked directly to the climates of different areas. For example, a heat pump in Dallas Texas might save 50% in energy costs while a heat pump in Minneapolis Minnesota might only save 20%. In a warm place, like Miami, the savings might hit 60%, or more. These numbers are significant when you consider that even a 20% savings in energy costs over the course of a 30-year mortgage mount up.

As an example, you can figure out how much one might save if a heat pump saved 20% in operating cost for 30 years, given an annual energy expense of $1,500. Assuming energy costs remained static for 30 years, how much would a 20% savings amount to? The answer is $9,000. How about if the savings were 50%, which is a reasonable amount for homes in many locations to expect. The savings would be $22,500.

As you can see by the above examples, heat pumps can pay for themselves in most any climate. The payback period will vary with the climate, but if you stay in the same house long enough, the additional initial cost of the heat pump can be recovered.

Interior temperatures

Heat pumps work best when they are not asked to handle extreme shifts in cold temperatures. For example, a heat pump in the home of a young couple who are comfortable at an interior temperature of 68 degrees F will provide more satisfactory results in a cold climate than it would in a home where an elderly couple might want a room temperature of 75 degrees F.

If you live in an area where the temperature drops below 20 degrees F for extended periods of time, you will have to plan on one of two things when buying a heat pump: buy a unit designed for cold climates, or become accustomed to cooler interior temperatures.

Because heat pumps in cold climates have to work harder to maintain warm interior temperatures, it is crucial that the home being heated is as energy efficient as possible. If there are drafts around windows and doors, the effectiveness of the heat pump will falter. Good insulation, tight windows and doors, and a modest room temperature will all help to ensure a better job from the heat pump.

When you are planning the insulation of the home, don't forget to insulate the duct work. Uninsulated ducts can have a detrimental effect on the performance of a heat pump. It is important to keep the air in the ducts as warm as possible until it is delivered through the heat registers.

Construction considerations

Construction considerations must also be taken into account when estimating the cost effectiveness of a heat pump. If a new house is being built, it can be built to standards that will allow the efficient use of a heat pump. Older houses that are being updated might not allow this same type of opportunity.

The last chapter briefly detailed remodeling homes that were not fitted with duct work or that had small duct work. When either of these situations arise, the cost for installing a heat pump can get out of hand, especially in cold climates where the cost savings on energy are marginal.

The expense that must be incurred to install duct work in a multi-story home that has no ducts, or inadequate duct work, can be enough to kill the deal. The following is an example of this type of situation.

Assume that you have just purchased a lovely, old farmhouse in Vermont. The house is a rambling two-story, with an ell addition that houses the kitchen and dining area. The house sits on a stone foundation with an open-air crawlspace under all but the back end, where there is an earthen cellar. Now that you own your dream retreat, you want to modernize it. Your plans call for replacing the old boiler and hot-water baseboard heat with an energy-efficient heat pump. Also, you don't want to rely on the wood stove that has been used to heat the kitchen and dining area, so you will be expanding the size of the heating system. Is the house a good candidate for a heat-pump conversion? Not likely.

As good as heat pumps are, they have their limitations, and the scenario you've just been given is one of them. Let's take the example apart, piece by piece, and see why a heat pump is not a logical choice.

The first problem you are faced with is an old house. That means insulation in the walls, attic, and crawlspace is minimal. This is not good for a heat pump. The house is in Vermont, a cold state, where the payback period will be long in coming. Strike two for the heat pump. Being an old farmhouse, the windows are probably loose fitting and uninsulated. Strike three, the heat pump's out, and we've still got several more strikes to come.

The house has no existing duct work. The combination of working in a crawlspace and chasing duct work into the upper level of the house will be very expensive. The open foundation will not protect the lower floors from cold. You can bet the electrical system will not be adequate for a heat pump, and there are probably other hidden obstacles.

To remodel this house to a point where a heat pump would make sense, the cost would be astronomical. Unless your original plans called for a major gutting and rehabbing of the house, a heat pump doesn't make sense.

It might be logical to use a one-piece heat pump to serve the kitchen and dining room, but it is highly unlikely that you could find any financial feasibility in converting the entire house to a heat-pump system.

If this same house had been located in a warm southern state, you might be able to find some justification in the long-term savings, but in Vermont, the payback is too little, too late. This example is just one other way of weighing the cost effectiveness of a heat pump.

A one-story house

A one-story house, with a full basement, could change the equation of the earlier example. Assume that you had just purchased a home that had all of the living space on one level and a full basement beneath it. Even if the insulation, electrical service, and windows and doors were still deficient, this home might well be worth the investment of a heat-pump conversion.

In all probability, you would want to modernize the energy-efficiency of the home regardless of the type of heating system you chose to use. This would mean blowing insulation into the exterior walls and attic. The windows could either be replaced with new, energy-efficient units or covered with storm windows.

Running duct work in the basement of a one-level house is not such a big deal. There is plenty of space to work, and the full foundation prevents unwanted drafts. If your plans called for an energy-efficiency updating of the home, you would be tending to the windows and insulation, regardless of the heating system.

Updating the electrical service might not have been in your plans, and that aspect is still a cost you would have to consider being related to the heat-pump conversion. However, if the electrical service was an old, 60-amp fused service, it should be upgraded whether you swap heating systems or not.

You might find that the only major cost associated with the heat-pump conversion, aside from the heating unit itself, was the installation of duct work. In prime working conditions, which this sample house offers, the duct work would not be an unmanageable expense. If you're handy, you could have the ducts fabricated for you in a metal shop, and you could install them yourself.

All of a sudden, the feasibility of converting an old home to a heat-pump system is not so financially stupid. In fact, the improvements might prove to be useful at tax time, and the added benefit of having air conditioning would not be resented. There really is much to consider when assessing the value of a heat pump.

Add-on units

Add-on units bring another dimension to the heat-pump question. If your home has an existing forced hot-air furnace, installing an add-on heat pump could make a lot of sense. By incorporating an add-on unit into your present hot-air duct work, you could improve the efficiency that you presently have at an affordable cost.

The add-on unit could work to save you money during the heating system and to cool your home in the hot months of the year. If the climate is such that the heat pump reaches its balance point (the point where it is no longer cost effective) the exiting furnace could kick in and warm your home during extreme temperatures. In many ways, this might be the best of both worlds for people in extremely cold climates.

Balance the scales

To balance the scales on the cost effectiveness of heat pumps, you have to do more than look at the price tag. As you have seen in this chapter, there is much more to consider than just the initial acquisition and installation cost. There are truly very few conditions that will not permit the cost-effective use of a heat pump in a home being constructed; however, there are times when remodeling and updating older homes that heat pumps just don't fill the bill.

If you do the necessary research, and invest your time before your money, you can determine, with confidence, whether or not a heat pump is right for you. The chances are good it will be.

To make a heat pump cost effective, you must have an efficient design. Most homeowners do not design their own heating and cooling systems, but that doesn't mean they shouldn't have a general working knowledge of what makes a practical design. An inefficient design can cripple the best heat pump, so design measures are important to the overall performance of any heating and cooling system.

3

Designing your heat-pump system

Designing your heat-pump system is crucial to its efficient operation. If the system is not designed properly, all sorts of problems can arise. Retrofitting an existing home with a heat pump allows fewer options in the design stage, but the importance of the design remains paramount.

When many people consider the design of a heat-pump system, they only think about the duct work. Certainly, duct work is a big part of the system, but it is far from being the only design consideration. There are issues involving everything from the placement of the outside unit to the location of the thermostat. Before an efficient design can be made, many factors must be weighed.

If you elect to hire a contractor to supply and install your heat pump, you might not have to be concerned about many of the design factors. There are, however, some elements of the design that will effect you, even if you are not doing the work yourself. For example, you might prefer one location over another for the placement of the outside unit. Your reason might be one of only appearance, but appearance is an important attribute of a home.

It helps to have at least a broad-brush understanding of how and why a heat pump is installed the way it is. This type of knowledge makes it easier for you to communicate clearly with your contractor, and clear communication is a key to successful negotiations and work.

As an average homeowner, you cannot possibly be expected to know all the technical jargon that is used within the heating ventilation and air-conditioning (HVAC) trade. You should, however, endeavor to learn enough about the types of heat pumps available to be able to discuss them competently with your contractor.

If you will be installing your own heat pump, it is necessary to extend your knowledge in all areas of design and installation procedures. When you consider that most people who design and install heat pumps have been trained professionally and have years of experience, you are taking on a large responsibility to do the entire job yourself.

You might find that the task of installing a new heat pump is more than you are comfortable with. Perhaps there are parts of the job you will do and other parts that you will contract out to professionals. In any case, there are some aspects of designing a good heat-pump system that you should become familiar with.

This chapter is intended to help you overcome any deficiencies you might have in understanding the design principles for heat-pump installations. The advice given here will cover water-source heat pumps, earth-source heat pumps, and air-source heat pumps. As with most things in life, proper planning is a major contributor to favorable results, and this chapter is going to help you plan the layout of your heat-pump system. The following explains design considerations for the three major types of heat pumps.

Air-source heat pumps

Air-source heat pumps are normally the least expensive type of heat pump to install. They are also the most prolific of the three major types of heat pumps.

Air-source heat pumps have limitations on their efficiency. If you live in a state that experiences extremely cold temperatures during the winter months, an air-source heat pump probably isn't the best heating solution for your home.

Most air-source heat pumps begin to lose their efficiency when the outside temperature drops below 28 degrees F. Once the outside temperature dips to 10 degrees F, air-source heat pumps are normally straining to maintain a comfortable temperature in the homes they serve. Most of these heat pumps have back-up heating elements to assist them when the outside temperatures are extremely cold, but the operation of the back-up heat drastically effects the efficiency of the heat pump.

Your first step in designing a heat-pump system for your home is to decide if an air-source heat pump is cost-effective in your location.

If your winter temperatures get very cold and stay there for extended periods, like they do in Maine, an air-source unit probably won't produce satisfactory results. Chapter 4 gives plenty of advice on shopping for the right type of heat pump.

Assuming that you have decided to install an air-source heat pump, you need to take several factors into consideration. Because almost all residential applications of heat pumps involve split systems, those are the ones covered here. A split system is one where one part of the heat pump is installed outside the home, and the other part of the heat pump is installed inside the home.

Before you can get too far into your design stage, you must know what size heat pump will be required for your home. Chapter 8 provides you with the information needed to size your unit.

Roof-top installations

Roof-top installations are common on commercial buildings, but rare in residential applications. There are several reasons why residential dwellings don't utilize roof-top installations.

Because most homes don't have flat roofs, setting the outside unit of a heat pump on the roof is rarely practical. Even if you have a flat roof that has the structural capacity to house an outside unit, the roof is still not the best place to put the equipment.

Outside units that are installed on roofs are subjected to blazing heat in the summer and cold winds in the winter. These two factors play a large role in the efficiency of a heat pump. To be at its best, the outside unit of a heat pump should be shaded in the summer and protected from wind in the winter. This goal is much easier to accomplish when the equipment is installed at ground level.

Ground-level installations

Ground-level installations of outside units for heat pumps far exceed roof-top installations when residential properties are involved. Not only are ground-level installations more convenient to work with, they also allow the heat pump to be more energy efficient.

There are several rules that should be obeyed when designing the location for an outside unit.

Clearance When you are deciding where to set the outside unit, you must keep the clearance requirements for the unit in mind. A good rule of thumb is to allow 30 inches of clearance, horizontally, in all directions from the coil of the outside unit. Vertically, a clearance of five feet will generally keep you out of trouble.

Wind protection Wind protection should be provided, in some form, for the outside unit. The house will protect the coil of the unit on at least one side. Many people plant low-growing bushes around their outside units to both improve the appearance of their home and to protect the coil from wind. If you decide to plant shrubbery as a wind screen, remember that many types of low-growing bushes grow out in width. Plant the bushes far enough away from the outside unit to maintain a clear distance of at least two feet, and preferably 30 inches.

Some soil conditions make planting natural wind screens difficult. If you can't accomplish the protection with shrubbery, you can build a decorative wind screen out of some other type of material, such as wood.

Overhangs The overhangs of roofs can cause problems for heat pumps that are installed under them. Avoid a location that will put the outside unit in the path of falling ice, snow, and rain that might come off the roof.

Snow Deep snow can bury the outside unit of a heat pump if the equipment is not elevated on a stable platform. While homeowners in southern states can set their outside units on concrete pads, close to the ground, residents in snow country must take accumulated snow build-up into consideration.

To keep snow from interfering with the clearance requirements of an outside unit, the unit must be elevated to a point above an anticipated snow accumulation. There are special stands available for setting outside units on, or a homemade stand can be used. Options for elevating the outside unit include the use of cinder blocks, concrete, angle iron, commercial stands, etc. The important thing is to make sure the stand is level and stable.

Drainage Drainage arrangements must be made for the outside unit. This is typically done by installing crushed stone and sand in the area surrounding the location of the outside unit. Chapter 10 gives you all the specifications needed for drainage and other aspects of actually setting the heat-pump pad into place.

The indoor unit

The indoor unit for an air-source heat pump can be installed in numerous locations. Basements are an ideal place for installing inside units, but closets, attics, laundry rooms, and crawl spaces can all be used to house the interior unit. Check the manufacturer's suggestions before making your installation, but any place that allows room to work on the unit should be suitable for the installation.

Close proximity The indoor unit should be installed in close proximity to the outdoor unit. The closer, the better. For example, if you were to set the outside unit so that it was near the kitchen, you should seek a location close to the kitchen for the inside unit.

Condensation piping Heat pumps remove moisture from homes, and this moisture is disposed of through condensation piping. The condensate line will typically drain into a portion of the plumbing system, such as a floor drain. A small trap (a U-bend in the piping) is often required on the condensate piping, as shown in Fig. 3-1. When you are designing the placement of the inside unit, make sure the equipment will be high enough above the floor to allow for the installation of a trap.

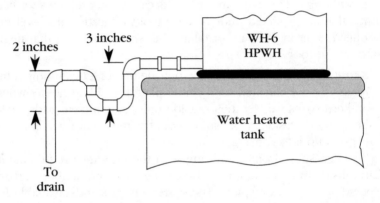

3-1 *Condensate trap.* Crispaire Corporation

A central location A central location in the home is the best place for the inside unit. This will allow for better air distribution. In other words, if you have the choice of putting the inside unit at one end of the home or near the middle of the home, choose the location in the middle.

Routing the refrigeration tubing

When routing the refrigeration tubing that connects the inside unit to the outside unit, you should avoid all sharp turns. When the tubing is run through a wall, it should be protected from chaffing with the use of a sleeve, such as a piece of plastic plumbing pipe. Plan on a sleeve that will be large enough to accommodate the tubing, the insulation used to wrap the tubing, and electrical wires that will be run with the tubing.

When planing the installation of the tubing, you should avoid situations that will require the tubing to come into contact with floor joists and wall studs. Plan to use hangers that will keep the tubing from making direct contact with structural members. This is done to reduce noise in the home when the units are running.

Oil

Oil must be available to the compressor of a heat pump. When inside and outside units are installed at different heights above or below ground level, special arrangements might be needed to ensure the proper conveyance of oil. Read the manufacturer's recommendations to see if and where oil traps might be needed.

Back-up heat

Back-up heat is needed with almost all air-source heat pumps. This is not much of a factor during the design stage, but make sure that the heat pump you buy has adequate arrangements for back-up heat (Fig. 3-2).

Duct work

Duct work is what carries the air from a heat pump to the register outlets. It is the duct work that often causes the most trouble in designing a heat-pump system for houses that have been built without heat pumps installed. When the new construction of a house is being planned, duct work is not such a bother.

Heat pumps often require larger duct work than forced hot-air furnaces. Many people who are replacing hot-air furnaces assume that their existing duct work can be used effectively with the new heat pump. While some of the duct work often can be used, much of it is usually too small.

When duct work is nonexistent in an existing home, installing it can be more than just a little troublesome. Chapter 6 talks about many aspects of installing duct work in existing homes.

A typical layout will have a trunk line (the main duct system) running along the length of the home. The size of the duct work will get smaller as it gets longer. Trunk lines often run under the middle of homes. Most older trunk lines are formed from sheet metal.

Supply ducts are cut into the trunk line and extended to the boots of floor registers. The supply ducts are frequently made of round, flexible duct material, and they are easy to work with. There is also a need for return-air duct work. Chapter 12 deals with all aspects of supply and return ducts.

Optional Equipment

Supplementary Heaters

UNIT MODEL	ELECTRIC HEATER MODEL	RATED VOLTAGE	PHASE	HEATER CAPACITY	
				KW	BTUH
WCC018F1	BAYHTRN105A	208/240	1	3 74/4 98	12800/17000
	BAYHTRN108A	208/240	1	5 76/7 68	19700/26200
WCC024F1	BAYHTRN105A	208/240	1	3 74/4 98	12800/17000
	BAYHTRN108A	208/240	1	5 76/7 68	19700/26200
	BAYHTRN110A	208/240	1	7 47/9 96	25500/34000
	BAYHTRN112A	208/240	1	8 64/11 52	29500/39300
WCC030F1 WCC036F1 WCC042F1	BAYHTRN105A	208/240	1	3 74/4 98	12800/17000
	BAYHTRN108A	208/240	1	5 76/7 68	19700/26200
	BAYHTRN110A	208/240	1	7 47/9 96	25500/34000
	BAYHTRN112A	208/240	1	8 64/11 52	29500/39300
	BAYHTRN115A*	208/240	1	11 21/14 94	38300/51000
	BAYHTRN117A*	208/240	1	12 97/17 28	44200/59000
WCC048F1 WCC060F1	BAYHTRN110A	208/240	1	7 47/9 96	25500/34000
	BAYHTRN112A	208/240	1	8 64/11 52	29500/39300
	BAYHTRN115A*	208/240	1	11 21/14 94	38300/51000
	BAYHTRN117A*	208/240	1	12 97/17 28	44200/59000
	BAYHTRN123A*	208/240	1	17 28/23 04	59000/78600
WCC036F3	BAYHTRN310A	208/240	3	7 47/9 96	25500/34000
	BAYHTRN315A	208/240	3	11 18/14 90	38100/50800
	BAYHTRN310F	208/240	3	7 47/9 96	25500/34000
WCC042F3	BAYHTRN310A	208/240	3	7 47/9 96	25500/34000
	BAYHTRN315A	208/240	3	11 18/14 90	38100/50800
	BAYHTRN320A	208/240	3	14 94/19 92	51000/68000
	BAYHTRN310F	208/240	3	7 47/9 96	25500/34000
WCC048F3	BAYHTRN310A	208/240	3	7 47/9 96	25500/34000
	BAYHTRN315A	208/240	3	11 18/14 90	38100/50800
	BAYHTRN320A	208/240	3	14 94/19 92	51000/68000
	BAYHTRN310F	208/240	3	7 47/9 96	25500/34000
WCC060F3	BAYHTRN310A	208/240	3	7 47/9 96	25500/34000
	BAYHTRN315A	208/240	3	11 18/14 90	38100/50800
	BAYHTRN320A	208/240	3	14 94/19 92	51000/68000
	BAYHTRN330A*	208/240	3	22 36/29 80	76300/101700
	BAYHTRN310F	208/240	3	7 47/9 96	25500/34000
WCC036F4	BAYHTRN410A	480	3	9 96	34000
	BAYHTRN415A	480	3	14 90	50800
	BAYHTRN410F	480	3	9 96	34000
WCC048F4	BAYHTRN410A	480	3	9 96	34000
	BAYHTRN415A	480	3	14 90	50800
	BAYHTRN420A	480	3	19 92	68000
	BAYHTRN410F	480	3	9 96	34000
WCC060F4	BAYHTRN410A	480	3	9 96	34000
	BAYHTRN415A	480	3	14 90	50800
	BAYHTRN420A	480	3	19 92	68000
	BAYHTRN430A	480	3	29 80	101700
	BAYHTRN410F	480	3	9 96	34000
WCC036FW	BAYHTRNW10A	600	3	9 96	34000
	BAYHTRNW15A	600	3	14 90	50800
WCC048FW	BAYHTRNW10A	600	3	9 96	34000
	BAYHTRNW15A	600	3	14 90	50800
	BAYHTRNW20A	600	3	19 92	68000
WCC060FW	BAYHTRNW10A	600	3	9 96	34000
	BAYHTRNW15A	600	3	14 90	50800
	BAYHTRNW20A	600	3	19 92	68000
	BAYHTRNW30A	600	3	29 80	101700

NOTES 1 Any power supply and circuits must be wired and protected in accordance with local electrical codes
2 The MCA values listed are for electric heater only
3 Field wire must be rated at least 75 C

3-2 *Optional heat-pump equipment table.* The Trane Company, an American-Standard company

NO. OF STAGES	KW/STAGE		MCA (2)	MAX. FUSE OR HACR CKT BKR SIZE (4)	CANADA ONLY MAX CKT BKR. SIZE (5)
	1	2			
1	3.74/4.98	—	22/26 (3)	25/30	30/30
1	5.76/7.68	—	35/40 (3)	35/40	40/40
1	3.74/4.98	—	22/26 (3)	25/30	30/30
1	5.76/7.68	—	35/40 (3)	35/40	40/40
1	7.47/9.96	—	45/52 (3)	45/60	50/60
2	4.32/5.76	4.32/5.76	52/60 (3)	60/60	60/60
1	3.74/4.98	—	22/26 (3)	25/30	30/30
1	5.76/7.68	—	35/40 (3)	35/40	40/40
1	7.47/9.96	—	45/52 (3)	45/60	50/60
2	4.32/5.76	4.32/5.76	52/60 (3)	60/60	60/60
2	7.47/9.96	3.74/4.98	67/78 (3)	70/80	70/100
2	8.64/11.52	4.33/5.76	78/90 (3)	80/90	100/100
1	7.47/9.96	—	45/52(3)	45/60	50/60
2	4.32/5.76	4.32/5.76	52/60 (3)	60/60	60/60
2	7.47/9.96	3.74/4.98	67/78 (3)	70/80	70/100
2	8.64/11.52	4.33/5.76	78/90 (3)	80/90	100/100
2	8.64/11.52	8.64/11.52	104/120 (3)	110/125	125/125
1	7.47/9.96	—	26/30	30/30	30/30
1	11.18/14.90	—	39/45	40/45	40/50
2	3.74/4.98	3.74/4.98	26/30	30/30	30/30
1	7.47/9.96	—	26/30	30/30	30/30
1	11.18/14.90	—	39/45	40/45	40/50
2	7.47/9.96	7.47/9.96	53/60	60/60	60/60
2	3.74/4.98	3.74/4.98	26/30	30/30	30/30
1	7.47/9.96	—	26/30	30/30	30/30
1	11.18/14.90	—	39/45	40/45	40/50
2	7.47/9.96	7.47/9.96	53/60	60/60	60/60
2	3.74/4.98	3.74/4.98	26/30	30/30	30/30
1	7.47/9.96	—	26/30	30/30	30/30
1	11.18/14.90	—	39/45	40/45	40/50
2	7.47/9.96	7.47/9.96	53/60	60/60	60/60
2	11.18/14.90	11.18/14.90	78/90	80/90	100/100
2	3.74/4.98	3.74/4.98	26/30	30/30	30/30
1	9.96	—	15	15	15
1	14.90	—	22	25	30
2	4.98	4.98	15	15	15
1	9.96	—	15	15	15
1	14.90	—	22	25	30
2	9.96	9.96	30	30	30
2	4.98	4.98	15	15	15
1	9.96	—	15	15	15
1	14.90	—	22	25	30
2	9.96	9.96	30	30	30
2	14.90	14.90	45	45	50
2	4.98	4.98	15	15	15
1	9.96	—	13	15	15
1	14.90	—	18	20	20
1	9.96	—	13	15	15
1	14.90	—	18	20	20
2	9.96	9.96	24	25	30
1	9.96	—	13	15	15
1	14.90	—	18	20	20
2	9.96	9.96	24	25	30
2	14.90	14.90	36	40	40

4 The HACR circuit breaker is for U.S.A. installations only
5 For Canada installation reference only
* Heater uses fuses

From Dwg 21C729567 Rev 7

Typical Single Rack Heater

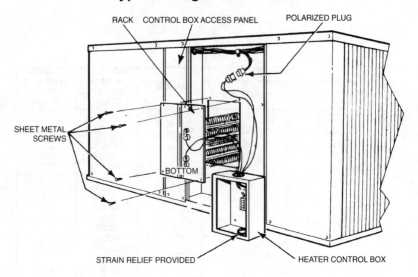

Typical Dual Rack Heater

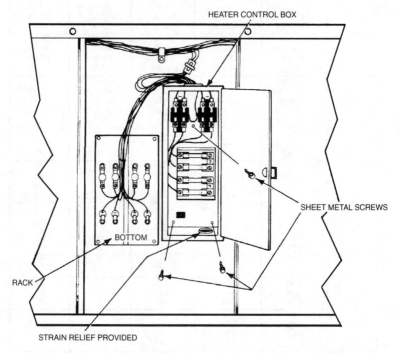

3-2 *Continued.*

During the design phase of your work, the two most important jobs you have pertaining to duct work are: sizing it correctly and finding suitable places to install it.

Chapter 6 discusses ways of installing duct work in existing homes. If you are building a new home, routing the duct work will not be much of a problem. By reviewing your blueprints, you can find all sorts of places that will allow the installation of the duct work. Depending on how complete your set of blueprints is, you might even have a page dedicated to showing the size and locations of duct work.

Computing the size of duct work is not complicated, if you have the proper charts and tables to work with. Most manufacturers of heat pumps will supply you with charts and tables for recommended sizes.

The tables and charts provide assistance in sizing both trunk lines and branch ducts (Tables 3-1 through 3-4). There are provisions for round ducts and rectangular ducts. By looking at how many cubic feet of air movement is required, you can go across the chart and find the proper duct sizes.

Table 3-1 Conversion table for residential ductwork (for 4-inch to 8-inch round branch ducts)

*CFM	Round duct size	Rectangular duct size
50	4 inch	4" × 4"
75	5 inch	4" × 5" & 4" × 6"
100	6 inch	4" × 8" & 5" × 6"
125	6 inch	4" × 8", 5" × 6", & 6" × 6"
150	7 inch	4" × 10", 5" × 8", & 6" × 6"
175	7 inch	5" × 10", 6" × 8", 4" × 14", & 7" × 7"
200	8 inch	5" × 10", 6" × 8", 4" × 14", & 7" × 7"
225	8 inch	5" × 12", 7" × 8", & 6" × 10"

*CFM = cubic feet per minute

Table 3-2 Conversion table for residential ductwork (10-inch to 14-inch round main or trunk ducts)

*CFM	Round duct size	Rectangular duct size
400	10 inch	4" × 20", 7" × 10", 6" × 12", & 8" × 9"
450	10 inch	5" × 20", 6" × 16", 9" × 10", & 8" × 12"
500	10 inch	10" × 10", 6" × 18", 8" × 12", & 7" × 14"
600	12 inch	6" × 20", 7" × 18", 8" × 16", & 10" × 12"
800	12 inch	8" × 18", 9" × 15", 10" × 14", & 12" × 12"
1000	14 inch	10" × 18", 12" × 14", & 8" × 24"

*CFM = cubic feet per minute

**Table 3-3 Conversion table for residential ductwork
(16-inch to 20-inch round main or trunk ducts)**

*CFM	Round duct size	Rectangular duct size
1200	16 inch	10" × 20", 12" × 18", & 14" × 15"
1400	16 inch	10" × 25", 12" × 20", 14" × 18", & 15" × 16"
1600	18 inch	10" × 30", 15" × 18", & 14" × 20"
1800	20 inch	10" × 35", 15" × 20", 16" × 19", 12" × 30", & 14" × 25"
2000	20 inch	10" × 40", 12" × 30", 15" × 25", & 18" × 20"

*CFM = cubic feet per minute

**Table 3-4 Conversion table for residential ductwork
(9-inch to 12-inch round branch ducts)**

*CFM	Round duct size	Rectangular duct size
250	9 inch	6" × 10", 8" × 8", & 4" × 16"
275	9 inch	4" × 20", 8" × 8", 7" × 10", 5" × 15", & 6" × 12"
300	10 inch	6" × 14", 8" × 10", 7" × 12"
350	10 inch	5" × 20", 6" × 16", & 9" × 10"
400	12 inch	6" × 18", 10" × 10", & 9" × 12"
450	12 inch	6" × 20", 8" × 14", 9" × 12", & 10" × 11"

*CFM = cubic feet per minute

Most suppliers of heat pumps will be happy to size your heating system for you. If you provide the suppliers with a set of blueprints, they can size the system and provide you with a material list of the items needed to get the job done. If you are a die-hard do-it-yourselfer, you can obtain sizing charts and tables from various manufacturers of heat pumps or from the American Society of Heating, Refrigeration, and Air Conditioning Engineers (ASHRAE). ASHRAE can be contacted by calling (404) 636-4404 or by writing to: ASHRAE, 1791 Tullie Circle NE, Atlanta, GA, 30329.

Electrical service

The use of a heat pump will require a 200-amp electrical service. Many older homes have 60-amp or 100-amp electrical services, and this is an expensive problem when converting to a heat pump. Having a licensed electrician upgrade an electrical service will cost quite a bit, so don't overlook this part of financial budgeting when designing your system.

An electrician will be needed to install disconnect boxes for the heat-pump units. Due to the dangers involved in working with electricity, most homeowners should hire licensed electricians to perform all phases of electrical work.

Water-source heat pumps

Water-source heat pumps are not as common as air-source heat pumps. One reason they are not so common is the expense of installing them. Another reason is that many consumers have never heard of them. Water-source heat pumps are probably the most cost-effective type of heat pump you can buy.

Most water-source heat pumps depend on well water for their operation. However, it is possible to use surface water, such as a pond or river, to provide the heat source for water-source heat pumps. One big advantage to well water is that it is fairly constant temperature. In many parts of the country, well water temperatures stay between 48 degrees F and 50 degrees F throughout the year. The exact temperature fluctuates some, and different states report different temperatures, but the water can be counted on to stay moderate all year.

The fact that well water stays so warm in winter is a big advantage when using a heat pump. For example, an air-source heat pump might have to pull its heat from outside air with a temperature well below freezing. A well-water heat pump, under the same conditions, will be drawing its heat from the much warmer well water. This results in less work for the heat pump and more cash savings for the user.

While there is little question that a water-source heat pump is more cost-effective to operate than an air-source heat pump, there are downsides to water-source systems. The biggest drawback is the cost of having the system installed.

When you set out to design an effective water-source system, you will run into many of the same considerations that must be dealt with when planning an air-source system. In addition to all the overlapping factors, there are many unique aspects of water-source systems to be evaluated.

The well

The first major design step for a heat pump that receives its heat from well water is the well itself. Do you have an existing well that can be used, or will you need to have one drilled? The expenses involved with drilling wells is one serious drawback to well-water heat-pump systems.

Not all wells are suitable for use as a heat source for a heat pump. Wells that are shallow, have a low reservoir of water, or a low flow rate will not produce satisfactory results when used as a heat source. Therefore, drilled wells are usually the only type of well that will provide dependable service.

The recovery rate of a drilled well must be known in order to properly plan the installation of a well-water heat pump. A professional well driller can test the flow rate of a well to determine how many gallons per minute (GPM) the well is capable of producing. Because the results of this test are crucial to the effective operation of the new heat pump, it is best to engage a trained professional to perform the test.

Dual duty

If you already have a drilled well that provides potable water to your home, the well might be capable of performing dual duty. By this, I mean, the well might be able to go on providing water for domestic uses as well as providing a heat source for your new heat pump.

When a well is called upon to perform double duty, it should have a flow rate capable of providing adequate water for both types of demand simultaneously. Not all wells have a strong enough GPM to do this, and there are ways to work around such a problem, but the ideal situation will not call for a compromise.

Reverse-acting

If you must compromise, a reverse-acting pressure switch is the most cost-effective way to control the water flow between domestic use and heat-pump use. The switch is installed in the water pipe between the heat pump and the pressure tank in the well system.

The switch can be programmed to cut off the heat pump when available water pressure drops to a preselected level. For example, if two showers were drawing water simultaneously, the switch might cut the heat pump off to allow enough water pressure to run both shower heads.

If the well system is equipped with an adequate pump and a pressure tank, as most well systems are, a reverse-acting pressure switch is a good compromise when needed.

An alternative

An alternative to a reverse-acting pressure switch is the installation of a stronger pump and a larger pressure tank. The combination of a pump with a higher GPM rating and a larger reserve of water in the pressure tank can overcome the need for a reverse-acting pressure switch.

If you opt for a larger pump, be sure that the pump will not pump water faster than the well can replenish it. For example, assume your well has a GPM rating of four and a pump with a GPM rating of three is installed. A pump with a GPM rating of five would have the ability to drain the well. Because the well only produces four gallons of water per minute, pumping water at a rate of five gallons per minute will not work for very long.

The advantage of a larger pump and pressure tank is that the heat pump will not have to shut down during periods of peak use. The disadvantage is the cost of the pump and the larger pressure tank. Peak demands on water for domestic use is typically minimal. Taking a shower or filling a washing machine are two times when the heat pump might have to cut off, but these activities don't take long or occur frequently throughout a 24-hour period. Therefore, a reverse-acting pressure switch often provides adequate service at a reduced cost.

Mineral content

If the well has a high mineral content, the effects of the well water can be detrimental to the heat-pump system. High concentrations of acid, iron, calcium, magnesium, and similar materials can reduce the life expectancy of a heat-pump system.

Many plumbing contractors and most water-conditioning companies will test your well water free of charge. They come to your home and take a sample of the water. Some companies do an in-home test of the water while you watch. Others send the water off to a lab to have it analyzed.

If there are high concentrations of corrosive materials in the water, water-conditioning equipment can be installed to neutralize the water. The cost of this equipment and its installation is substantial, so have your well water tested before you make a firm commitment to a water-source heat pump that will use the well water as a heat source.

Modern heat pumps are being made with materials that resist the effects of most water infected with impurities. Unless the presence of impurities in your well is unusually high, there will probably be no need for water-conditioning equipment to be installed. If your well does show potential problems when tested, consult with various manufacturers to see if any water treatment is needed. By providing manufacturers with your test results, they will be able to give you definitive answers on what, if any, type of conditioning equipment is needed.

Where does the water go?

After the heat pump has extracted the water's heat, the water returns to ground, normally (Fig. 3-3). It is common for a second well, a return well, to be drilled into the same vein of water that provides water to the supply well. However, this practice might not be acceptable in all states, because some states are more protective in their code restrictions than others. Before you commit to using a well-water heat pump, check with your local code enforcement office for details on any restrictions that might apply to their use or the use of return wells.

3-3
Return-well layout. Tetco

When a return well is drilled, it must be installed far enough away from the supply well so as not to create a thermal imbalance. The advantage to using well water as a heat source is its fairly constant temperature. If return water is dumped back into the aquifer that supplies the supply well, it is possible that the water being returned will lower the temperature of the well water in the supply well. This would defeat the primary advantage of a well-water heat pump.

Closed-loop heat pumps

If you want a water-source heat pump but don't want to (or can't) drill a well, you might consider using a closed-loop heat pump. Closed-loop systems consist of a series of pipes being installed below ground. The ground and the water in the pipes work to provide a heat pump with its needed heat source. Some people refer to closed-loop systems as water-source heat sources, and others call them earth-source heat sources.

Closed-loop systems are relatively new, and their design can become quite complicated (Figs. 3-4 and 3-5, Tables 3-5 through 3-7). Because there is a lot of exchange between temperatures in the ground around the loop, one must be careful not to allow the changes in temperature to become too drastic. This is usually accomplished by installing an expansive piping system.

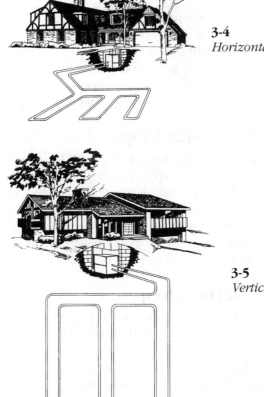

3-4
Horizontal closed loop. Tetco

3-5
Vertical closed loop. Tetco

Table 3-5 Fluid volume of pipe contents for copper pipe

Pipe size	Number of gallons
1 inch	4.1
1¼ inch	6.4
1½ inch	9.2

**Table 3-6 Fluid volume of pipe
contents for polybutylene pipe**

Pipe size	Number of gallons
¾ inch	2.8
1 inch	4.5
1¼ inch	7.8
1½ inch	11.5
2 inch	18

Table 3-7 Projected flow-rates for various pipe sizes

Pipe size	Projected flow rate
½ inch	2–5 gallons per minute
¾ inch	5–10 gallons per minute
1 inch	10–20 gallons per minute
1¼ inch	20–30 gallons per minute
1½ inch	30–40 gallons per minute

Horizontal loops

Horizontal loops of piping are often used with closed-loop systems. The layout of the piping is done at depths below the local frost line. It is often recommended that the loop be installed at a depth of six feet. The overall area of ground needed to install a horizontal loop varies. Many conditions contribute to the size and design of the loop. Soil conditions, temperatures, and other factors must be considered when engineering the loop.

Vertical loops

Vertical loops can be used in place of horizontal loops when a limited amount of surface area is available. With a vertical loop, the piping is buried in the ground vertically. Due to the depths reached with vertical loops, the likelihood of warmer soils being present is good, and this is better for the efficiency of the heat pump.

Both horizontal and vertical closed-loop systems require a special pumping kit. The kit is normally offered by manufacturers of heat pumps as an option accessory item.

Antifreeze is used in both types of closed-loop systems, but less of it is needed with vertical loops. Because vertical loops typically extend into warmer soil, less antifreeze is needed to pull the heat source.

Where to install the closed loop?

Where is the best place to install a closed loop? Some places are better than others for the installation of closed loops, especially if a horizontal design is used. Here are some creative places to put your piping.

A greenhouse

If you are planning to build a greenhouse at anytime in the future, you might consider installing a horizontal loop system where the greenhouse can be built over it. The greenhouse will collect a lot of heat that can be transferred to the loop, through the earth. In the hot months of summer, the greenhouse will have to be shaded and vented to allow the heat pump to function at its best.

If you decide on this approach, care must be taken not to damage the underground loop when constructing the greenhouse. Additionally, if problems ever arise with the underground piping, getting to the loop components for repair work will be difficult. A portable greenhouse, one with a metal frame and a plastic cover, is an ideal solution to this problem.

A vegetable garden

Burying the closed loop beneath a large vegetable garden is a great idea. In summer, the foliage from the garden plants will help shade the ground over the loop, keeping it cooler. In winter, the plants will be gone and the bare ground will help the loop to absorb heat from the sun's rays.

Near a septic field

Putting a horizontal loop near a septic field might not seem like a good idea at first, but it can actually be a very good location for the coil. Waste leaving a septic tank and entering a drainfield contains heat. As the waste water perks into the earth, the heat is given off. If you have your coils in the proximity of the distribution lines of a septic field, there is a good possibility for extra heat gain.

Locations to avoid

Just as there are some locations that provide good installation traits for a closed-loop system, there are some locations to avoid. Because the ground coil is made of pipes, there is a possibility for breakage and leaks. This possibility is escalated in some installation sites.

When you are planning the location of a horizontal coil, you should avoid areas that might be encroached upon by the roots of trees. Just as tree roots will invade sewer pipes, clogging and breaking them, they can also invade the piping in an underground loop.

Don't install the underground loop in an area that will be subject to traffic from heavy vehicles. The movement of vehicles over the piping could result in stress on the pipe and ultimately broken pipes.

If an underground loop ever develops a leak, finding and fixing the leak can be very difficult, even under the best of conditions. If the loop happens to be installed under a permanent structure, such as a house, the job can become impractical, if not impossible.

Would anyone really build a house over their underground coil? I've read where the practice is used to take advantage of the natural heat loss present under a home, but I've never seen it done. And, I wouldn't recommend doing it.

Locations for vertical loops are less troublesome. Because the pipe, or pipes, are installed vertically, much like a drilled well, there is less risk of damage to the piping. The vertical installation is less likely to be damaged by ground movement.

Most of the other design considerations for water-source and earth-source heat pumps are similar to those explained for air-source heat pumps. There are some installation differences, which are covered in the following chapters.

4

Shopping for a new heat-pump system

Shopping for a new heat pump can get confusing. There are so many options available that you might not have any idea where to begin. You could leave the decision of what type of heat pump to buy up a mechanical contractor who you hire to do the job. However, some contractors deal in better equipment than others do, and not all contractors are equal in their knowledge of heating and cooling systems. Do you really want to trust an unknown contractor to provide you with the best heat pump for your home?

If you are going to install your own heat pump, you might have to rely on the advice of salespeople. Most of these people have never installed a heat pump, thus they could not provide you with the facts needed to make a wise buying decision for your home.

When you begin to seriously consider the purchase of a heat pump, it is easy to see that if you don't know much about heating and cooling equipment, you might wind up with a less-than-perfect heat pump. So, what should you do to ensure that you get exactly what you want and need? Read this chapter carefully; it gives you the data needed to make a smart decision on what type of heat pump to buy.

Before you go shopping for a new heat pump, there are some aspects of the equipment you should already be familiar with. For example, what size heat pump are you looking for. Do you need a 1-ton unit or a 2-ton unit? Will you be looking for an air-to-air system or an air-to-water system? Will the heat pump you buy be a one-piece heat pump or a two-stage heat pump? Are you interested in a heat pump with a rectangular outside unit or an circular one? Does the shape of the outside unit make any real difference? These are just some of the questions that will come up when you begin to look for a heat pump.

Until you understand your needs for a heating and cooling system, you shouldn't go shopping. You might just become confused and frustrated. It is best to do your homework first, and then go window shopping. Once you have narrowed the field of competitive heat pumps, you can begin to talk seriously with sales representatives to make your final decision.

To prepare you for the myriad of choices, here is a step-by-step tour of important considerations that might effect your buying decision. To begin our trek into the purchasing possibilities for heat pumps, here is a review of the major types that are available.

One-piece heat pumps

One-piece heat pumps are normally used when only a relatively small area of living space is to be climate controlled. These units are fine for motel rooms and home additions. They are not well suited to heating and cooling whole houses. A one-piece heat pump is fine if you want to beef up the heating and cooling in a specific part of your home, but they will not serve you well as your primary source of heating and cooling for the entire home.

Two-piece heat pumps

Two-piece heat pumps, often called two-stage heat pumps, are by far the most popular style for most heating and cooling needs. One piece of the system is installed outside the home; this section of the equipment is referred to as the outside unit or the condenser unit. The second piece of equipment is installed inside the home. This unit doesn't require a lot of room and can be concealed in a closet. For an average home, a two-piece heat pump will be the equipment style of choice.

Air-to-air heat pumps

Air-to-air heat pumps are the most common type of heat pump used in residential applications. This is not to say they are the best type available, but they are the most popular.

Air-to-air heat pumps derive their heat source from the atmosphere. The outside unit pulls in warmth from the outside air and delivers it to the inside unit. Then the inside unit distributes the warmth throughout the home via duct work. This type of heat source, like any type, has its advantages and its disadvantages.

Advantages

There are several advantages to an air-to-air heat pump. First, air is available anywhere people live normal lives. This makes the source of heat for an air-to-air heat pump abundant. Also, air is considered to be a good source of heat for a heat pump.

Cost is another advantage. An air-to-air heat pump is the least expensive of all types of heat-source heat pumps. When you are evaluating the initial cost of a new heating and cooling system, you will be hard pressed to find a better value.

A final advantage is not one to be overlooked. Air-to-air heat pumps have been around a long time. This has allowed manufacturers to work the glitches out of them; so air-to-air units are typically dependable and long lasting.

Disadvantages

There are a few disadvantages to air-to-air heat pumps. The operation costs for an air-to-air heat pump are higher than those for other types of heat pumps. While some of the other types of heat pumps cost more to acquire, their operating expenses might be low enough to warrant the higher acquisition cost.

Frosting can be a problem with an air-to-air heat pump. In moist, cold temperatures, the outside unit will frost. This not only requires the unit to run frequently in its defrost cycle (increasing operating costs), it tends to make the heating unit less efficient in the production of warm air.

People who live in areas where the winter temperatures stay below 20 degrees F for extended periods of time might find that an air-to-air heat pump is not as economical as they thought it would be. If frequent snowfall accompanies the low temperatures, even more cost effectiveness will be lost.

During extreme cold, heat pumps can be made to switch automatically to a back-up heat source, such as electric resistance heating in the heat-pump system. This will bring the inside temperature of a home up to a comfortable level, but the operating costs will be substantial.

For a heat pump to work at its best, the difference in the temperature desired inside a home and the temperature of the heat source should be as minimal as possible. If you want your home to be a warm 72 degrees F on a day when the outside temperature is three degrees F, you are asking a lot of your air-to-air heat pump.

Of course, if you live in Florida, you don't have to be concerned with subzero temperatures. In fact, many parts of the United States have moderate enough temperatures for the consideration of back-up heat to be nonexistent. You must take your location into consideration when planning the purchase of any type of heat pump.

Air-to-ground heat pumps

Air-to-ground heat pumps are not as well known as air-to-air heat pumps, but there does seem to be a growing interest in them. When you are thinking about the different types of heat pumps, remember that the closer the temperature of the heat source is to the desired inside temperature of the home, the better the heat pump will work. This fact is one of the most desirable aspects of an air-to-ground system.

Earth-source heat pumps pull their heat from the earth. Piping coils are installed a foot or two below the frost line. A frost line is the depth in the earth that winter frost will penetrate. The depth varies from state to state and region to region. For example, where I used to live in Virginia, the frost line was 18 inches. Where I live now in Maine, the frost line is four feet, over twice the depth of that in Virginia.

By installing the coils well below the frost line, it is possible for an air-to-ground heat pump to enjoy a fairly constant temperature from its heat source. Unlike an air-source system, an earth-source system will normally have a heat source with a stable temperature of around 50 degrees F. This means that during cold weather, an earth-source heat pump will not have to work as hard as an air-source heat pump.

To illustrate the advantage that an earth-source system holds over an air-source system, here are two examples. For convenience, I used heat pumps for homes in Virginia and Maine, because those are the areas I have the most experience with.

The Virginia house

The Virginia house we will talk about has an air-to-air heat pump in the first phase of our illustration. During the winter, the inside temperature of this house is kept at 70 degrees F, not an unusually warm temperature. Much of the winter in the part of Virginia where the house is located sees daytime highs in the thirties and occasionally the forties. Nighttime lows linger in the twenties and thirties. A week or two of each winter surrounds the house with nighttime lows in the single digits, but almost never below zero.

The air-to-air heat pump is equipped with electric back-up heat, and it provides the warmth desired in the home. It does, however, have to run its expensive electric heat to keep the house warm on the colder nights. The operation of the electric heat makes the electric bill rise at the same time it is bringing the inside temperature up. This home benefit from the installation of an air-to-ground heat pump.

The air-to-air heat pump is maintaining an inside temperature of 70 degrees F. To do this, it is pulling from a daytime heat source that is typically in the thirties. For the sake of our comparison, say the air temperature during the day is 35 degrees F. Also assume the nighttime temperature averages 25 degrees F. This means that the spread between the inside temperature and the heat source during the day is 35 degrees F. At night, the spread is 45 degrees. Now, what would the spread be if a an earth-source heat pump was being used.

A properly installed earth-source heat pumps pulls from a heat source where the temperature is normally 50 degrees F. The daytime spread for the house in our example would be 20 degrees. This is 15 degrees warmer than the heat source for an air-to-air heat pump. At night, the spread for an earth-source system is still only 20 degrees F. The heat-source temperature for an air-to ground system doesn't fluctuate with the cycle of days and nights. This amounts to a nighttime heat-source temperature that is 25 degrees F warmer than the air-to-air system in the example.

It takes much less effort on the part of the heat pump to keep the house warm when it is getting its heat source from the earth. To expand on this, here's an abbreviated version of the example for my home in Maine.

The daytime highs for the last few weeks have been below 10 degrees F. Nighttime temperatures have been well below zero, with temperatures of twenty-below not unusual. This means that an earth-source system would have a 40-degree advantage during the day and a 70-degree advantage at night. It should be obvious why an earth-source system provides more consistent performance and lower operating costs than an air-to-air system does. Now let's run down the same list of advantages and disadvantages that we used for air-to-air systems.

Advantages

One of the best advantages of an earth-source system has just been explained, but there are others. For example, just as air is an abundant resource for air-to-air systems, the earth is under the feet of people most of the time. In other words, there is no shortage of availability for earth as a heat source. Additionally, operating costs for an air-to-ground system tend to be low.

Disadvantages

Money is the major disadvantage of an earth-source system. Air-to-ground heat pumps cost more to buy than air-to-air heat pumps. Not only do they cost more to purchase, they are more expensive to install.

Aside from just the issue of money, a leak in an underground system can be very difficult, and expensive, to locate and correct.

Earth-source systems are not as difficult to install during the construction of a home. After a house has been built and lived in, the excavation work needed to bury the coils can damage existing water pipes, gas pipes, electrical wires, and septic systems, not to mention the landscaping.

A last factor to consider when thinking of earth-source systems is their relative youth in the heat-pump industry. Because these systems are newer, and less tested, than air-to-air systems, there might be more bugs to work out of the equipment before it is as dependable as the old stand-by of air-to-air.

Well-water heat pumps

Well-water heat pumps are probably the most efficient heat pumps available. If we go back to the fact that the temperature of the heat source is relative to the heating ability of a heat pump, you can hardly beat a well-water system.

Well water, like the earth, maintains a fairly constant temperature throughout the year. In fact, it often mirrors the temperature of the earth in many parts of the country. In such cases, a well-water system will perform with about the same effectiveness of an earth-source system. There are, however, many parts of the United States where the temperature of well water rises well above 50 degrees F. This is when a well-water system leads the pack in efficiency.

Going back to the earlier example of the houses in Maine and Virginia, we can draw a comparison between a well-water system and an earth-source system. Well water in the part of Maine where I live runs at a temperature of approximately 48 degrees F. There is no real temperature advantage between a well-water system and an earth-source system. In Virginia, where I used to live, the well water runs at a temperature of around 60 degrees. This is a full 10 degrees warmer than the estimated temperature of an earth heat source in the same area. Based on these facts, a well-water system would have a constant 10-degree advantage over an earth-source system in central Virginia.

The temperature advantage for a well-water system in states where cold winter nights are common will range from six to 10 degrees. In extremely cold states, the earth-source heat pump might have a warmth advantage over the well-water system by two to six degrees. As you can see, it is necessary to refine your personal needs based on where you live.

Advantages

One of the biggest advantages to a well-water system is the low cost of operating the system. As long as the well supplying the system has a constant reserve of water, the stability of the heat source is good. Well-water heat pumps have proven themselves over time, so there are not many kinks to get out of the works.

Disadvantages

There are some disadvantages to a well-water heat pump. Unlike earth and air, well water is not available in all areas of the United States, but it can be found in more than three-fourths of the country. The fact that well water is not available in 100% of the country counts as a minor disadvantage.

The initial cost of a well-source heat pump will be more than that of an air-to-air system, and about the same as that of an earth-source system. However, the installation cost for a well-source heat pump might be less than that of most earth-source systems. The cost of drilling a well varies, but it is never cheap.

Some of the problems that might occur with a well-source heat pump are severe. The well could run dry, causing the heat pump to fail. Mineral deposits left in the equipment from the well water could gum up the works. These are factors that don't come into play with an air-to-air or air-to-ground system.

An alternative

There is an alternative to well-water systems that still used water as a heat source. These systems pull their water from ponds, lakes, and rivers. This type of system certainly isn't for everyone, but if you have access to suitable surface water, you might want to consider such a system.

Operating costs for a surface-water system are usually low, and the total cost of installation is usually second only to air-to-air systems. The stability of the heat source is better than that of an air-source system, but not as good as a well-water source or an earth-source.

Solar heat source

Have you considered a solar heat source for your new heat pump? Solar heat-pump systems are continuing to be perfected, and they show a lot of promise for the future. They are, however, expensive, perhaps too expensive to be feasible in many homes.

Most people have some idea how solar collection and storage works for heating water and homes. If you've driven down a neighborhood street and seen homes with large, glass-looking panels on their roofs, you can bet the home is using solar power. Other, less noticeable, uses of solar energy occur in homes equipped with sunrooms, large areas of glass in the main walls of the home, and so forth. Tile on the floors of homes with a lot of glass in the walls helps to catch and store the warmth from the sun. With these principles accepted as common place, it stands to reason that solar-powered heat pumps are not too far away from becoming popular. The catch is in figuring out how to make the installation of such systems affordable.

Advantages

The availability of solar power is very good. So is its suitability for use with heat pumps. With a properly installed solar-source heat pump, operating costs can be kept very low. These are all certainly good reasons to look to the sun for low-cost energy.

Disadvantages

While it is true that the sun will provide plenty of low-cost energy as a heat source, the expense of developing a system to harness that energy is not so agreeable. Because solar power is not always dependable, such as on cloudy days and at night, storage of the energy collected is a must. The price of solar panels and storage requirements needed to use the sun as a heat source for a heat pump can be a real eye-opener.

There are many factors that come into play when considering a solar-source heat pump. You must take into consideration the orientation of your home to the heat source. Obviously, an unobstructed southerly exposure is best. If your home sits in a valley, surrounded by mountains and tall trees, solar power is not going to be as effective as it could be for homes in other locations.

A house that is being built can be adapted to use solar power with less trouble than doing the same job on an older home. Remodeling an existing home to capitalize on solar power can become quite expensive.

Before you get too excited about cashing in on the warmth of the sun, talk to some qualified contractors in your area. They can help you determine the financial feasibility of a solar conversion or installation.

Coil material

The coil material in a heat pump can effect your operating costs. Coils in heat pumps look like a large mass of metal fins. These fins are usually made of aluminum, and they will bend easily. Because the fins play a crucial role in the effective operation of a heat pump, it is important that they be protected from abuse.

Coil fins attach to tubing to effectively expose the tubing to more air temperature. Copper tubing provides the best efficiency as a coil material, so look for a unit that is fitted with a copper coil.

Shape versus efficiency

Would you rather have a heat pump that looks good or one that works good? There are a lot of round outside units installed as part of two-stage heat pump systems. Seeing so many round units might give you the idea that a round unit is the best unit. This, however, is not necessarily true. In fact, round units are generally not the most efficient design.

Some people tend to buy products for reasons other than reliability and performance. A practical car could sit on a sales lot until it became an antique, but a pretty car will be sold before the first layer of dust settles on it. Because it is unusual to get something that is both pretty and practical, it seems most people opt for the pretty. This is the case, to some extent, with heat pumps.

Outside units that are round normally appeal to more people than rectangular units do. Does the fact that a round outside unit has more curb appeal make it a wiser choice? No, it doesn't. Can it be said then that rectangular units are better than round units? No, rectangular units might not be the most efficient. There is, however, a way to tell what units are the most efficient.

Earlier I talked about two ratings that give performance information on heat pumps. One was the coefficient of performance (COP) and the other was an energy efficiency ratio (EER). Check these two ratings to determine the effectiveness of prospective heat pumps. With both the COP and the EER, the higher the ratio rating is, the better the heat pump is.

What size do I need?

The first thought of many homeowners when they are contemplating the purchase of a heat pump is one of size. This is a question of importance. A heat pump of the wrong size is a very bad investment.

The design of heat pumps was detailed in chapter 3. Chapter 7 explains heat gain and heat loss. However, because we are talking about sizes, let's look at some simple, rule-of-thumb suggestions you might be interested in.

Heat pumps are sized in terms of tons, such as a 1-ton unit, a 2-ton unit, and so on. For each ton of size, a heat pump can remove 12,000 British Thermal Units (BTUs) per hour. That's right remove it, not provide it. This is a point that confuses many people.

When you think about buying a heating system, you don't normally think about removing heat, you think about creating it. Heat pumps, however, are sized based on their ability to remove heat, because they are almost always air conditioners as well as heaters.

A 1-ton unit will remove 12,000 BTUs. A 2-ton unit will remove 24,000 BTUs, and a 4-ton unit will remove 48,000 BTUs. The process of establishing a heat-gain-heat-loss rating is somewhat complicated, and it is detailed in chapter 7. For now, it is enough for you to know that heat pumps are sized in terms of tons, and that most homes will require a heat pump sized somewhere between two and five tons.

Getting advice

If you've never shopped for a heat pump before, you might be wondering who you should talk to. Most people think of talking to sales representatives in stores that sell heat pumps. These are certainly good people to talk to, but they are not the only ones who can help you. If you're thinking that you should talk to Heating-Ventilation-and-Air-Conditioning (HVAC) companies, you are correct. But, those two groups of professionals are not the only people to consider when shopping for a new heating and cooling system.

Homeowners

Other homeowners are great people to talk to when trying to make a buying decision for a new heat pump. Talk to your friends and neighbors. If these people have heat pumps, you will get some of the most honest feedback you are going to get from any source. If people you know and trust recommend a particular brand, especially if the brand name comes up more than once or twice, you can feel good about buying that brand. On the other hand, if a certain brand gets bad reviews from present owners, you might want to shy away from it. The benefit of talking to people who have first-hand experience as a consumer is enormous.

If you are new to an area and don't have any local friends, there is still a way to talk to people without knocking on the doors of strangers. Have contractors give you bids for the work, and request references from them. Ask specifically for references who have purchased equipment identical to the type you are contemplating. Make sure you get the names of customers who have used their heat pumps through a full season of heating and cooling. Then, you can call those homeowners and see if they are satisfied.

Utility companies

Call your local utility company and ask them for any literature or assistance they are willing to provide you with. The utility company will probably be happy to give you all sorts of help. For example, when I was building so many houses in Virginia, the utility company would perform the heat-gain-heat-loss calculations for me at no charge. They didn't just do this because I was building 60 homes a year, they would do it for anyone who asked.

Manufacturers

Manufacturers are often more willing to provide potential customers with detailed specifications than contractors or sales outlets are. I have never figured out why suppliers and contractors are hesitant to provide details on their products, but a lot of them are. If you want more information than you are able to get locally, call or write the manufacturer.

If you are wondering how you can locate the address or phone number of a manufacturer when all you know is the name of the equipment, a trip to your local library will solve your problem. Go to the reference desk and explain that you want to review directories that list U.S. manufacturers. There will be books there to help you.

When you are shown to the directories, you will have two easy choices. If you know the name of the company you want to contact, you can use one type of directory. When you want to contact several manufacturers, even though you don't know the company names, you can use the second type of directory.

If you don't know the company names, look them up under the heading of "Pumps, Heat." The heading might vary with different directories, but that's the heading I've always found them under. There will be a list of company names and addresses, along with the types of products they offer. Once you have the names, you can use the other directory to gain more information on the company. The second directory will have the companies listed, by name, in alphabetical order.

Dealing with suppliers

Dealing with suppliers can get tricky. You don't find heat pumps sitting around in department stores. Many of the suppliers who handle heat pumps won't sell them directly to consumers. They often sell only to licensed contractors at wholesale prices. This forces you, the homeowner, to deal with a contractor or retailer.

Because contractors and retailers usually handle only a limited number of brand names, they are going to try to sell you on the brands they have to offer. This puts you at a disadvantage, because you are not able to see the full spectrum of products available.

Contact your friends first. Their experience with particular brands can have a lot of influence on your buying decision. Also contact all the heat pump manufacturers you can find and request their product literature.

If you ask for an owner's manual for a particular model of heat pump, the manufacturer can supply it. There might be a nominal charge for the booklet, but once you've narrowed the field of heat pumps to only a few, the owner's manual is well worth buying.

Once you have the manuals, with complete specifications, you can compare heat pumps on an apples-to-apples basis. Product brochures only give you pretty pictures and sales hype. Owner's manuals give you the straight facts that should be used to base your decision on.

Local availability

Local availability of a particular brand of heat pump should be a consideration when looking for a new heating and cooling system. If the brand that checks out best in your research is not available within a 50-mile radius, you might want to move down your list to a second choice.

Whether you are installing the unit yourself or having a contractor install it for you, it is helpful to have reasonable access to the heat pump and to repair parts for it. The repair parts can become very important if you lose your heat in the middle of the winter and find that it will take six weeks to get a needed part.

Research

Research is the key to success in many aspects of life, including the purchase of a heat pump. By spending adequate time researching your options, you will not become confused and pressured when you talk with sales representatives. You will know what you want by make and model number, and there will be very little chance of getting talked into a heat pump that you don't want.

Too many people start their shopping procedure by talking with salespeople. Once the sales pressure is cranked up, a lot of the consumers follow the advice of someone who might know nothing more about heat pumps than what a sales manager has taught during weekly sales classes. Knowing what sales professionals call "features and benefits" is not the same has knowing product specifications. Features and benefits are what sell most items, but they are not always the best barometer of what makes a good heat pump. Figures 4-1 through 4-13 and Tables 4-1 through 4-22 provide you with examples, some right out of manufacturer's packages, of what you can expect to see when evaluating heat pumps.

4-1
Heat pump. Tetco

4-2
Heat pump. Tetco

4-3
Heat pump. The Trane Company, an American-Standard Company

4-4
Heat pump. The Trane Company, an American-Standard Company

4-5
Heat pump. The Trane Company, an
American-Standard Company

4-6 *Heat pump.* Friedrich Air Conditioning Co.

4-7
Outside uniTable. Bard Manufacturing Company

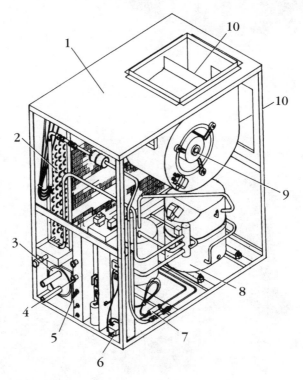

4-8 *Cross-section of a heat pump.* Tetco

USV Series

Vertical-Upflow-Cased Evaporator Coil
1-1/2 through 5 ton

4-9 Sample of manufacturer's technical data. Telco

EXPANSION VALVE KIT (field installed)

KIT NUMBER	FITS
978-9	1.5 - 2.0 TON
978-10	2.5 - 3.0 TON
978-11	3.5 - 4.0 TON
978-12	5.0 TON

Notes:
(1) Valves mount externally on both the USV and USH boxes
(2) Valves are external equalizing internal bleed type **(non heat pump)**

SIDE VIEW

FRONT VIEW

COIL CONNECTIONS

DRAIN CONNECTIONS

DIMENSIONS-PERFORMANCE

MODEL	NOMINAL TONNAGE RANGE	W	D	H	OUTLET W₀	OUTLET D₀	SUCTION O.D.	LIQUID O.D.	RETURN AIR OPENING SIZES — BREAK-AWAY WIDTHS (2) (BOTTOM OF COIL BOX)
USV324AP	1.5-2.0	16-1/2	21	13	14-1/2	17	5/8	3/8	9-1/4,10-3/4,12-7/8,14-3/8,15
DAX 2.0	2.0-2.5	21	21	17	19	17	5/8	3/8	10-3/4,12-7/8,15,16-1/4,17-1/2, 18,19-5/8,21
DAX 3.0	2.5-3.0	21	21	17	19	17	3/4	3/8	
USV342AP	3.0-3.5	24-1/2	21	21	22-1/2	17	3/4	3/8	12-7/8,15,16-1/4,17-1/2,18,19-5/8, 21,21-5/8,23-1/2,24-1/2
DAX 4.0	3.5-4.0	24-1/2	21	21	22-1/2	17	7/8	1/2	
DAX 5.0	4.0-5.0	28	21	22-1/2	26	17	7/8	1/2	15,16-1/2,21,21-5/8,23-1/2,24-1/2,25-1/2

Notes:
(1) Nominal capacities are based on 80DB/67WB at approximately 45 deg. suction with 400 cfm/ton.
(2) Bottom has multiple break-aways to allow matching with different furnaces.
(3) All refrigerant line connections are sweat.
(4) See current ARI Directory for certified performance data with selected manufacturers.

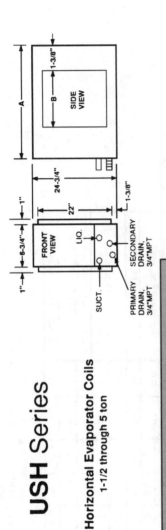

USH Series

Horizontal Evaporator Coils
1-1/2 through 5 ton

DIMENSIONS - PERFORMANCE					
MODEL	NOMINAL TONNAGE RANGE	A	B	SUCTION O.D.	LIQUID O.D.
USH324AP	1.5-2.0	22-1/2	18	5/8	3/8
DAH 2.0[3]	2.0-2.5	22-1/2	18	5/8	3/8
DAH 3.0[1]	2.5-3.0	22-1/2	18	7/8	3/8
USH342AP	3.0-3.5	30-1/2	26	7/8	3/8
DAH 4.0	3.5-4.0	30-1/2	26	7/8	3/8
DAH 5.0	4.0-5.0	37-1/2	33	7/8	1/2

See Notes 1, 3 and 4 above

Catalog No. USC292 (replaces USC1091)

All technical specifications subject to change without notice.

U.S. A/C PRODUCTS

4-10 Another sample of manufacturer's technical data. Tetco

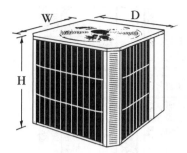

Basic model no.	"W" width	"D" depth	"H" height
24UHPQB 30UHPQB 36UHPQB	32½"	32½"	26"
42UHPQB 48UHPQB	32½"	32½"	26"
60UHPQB	32½"	32½"	36"

4-11 *Example of heat pump dimensions.* Bard Manufacturing Company

TABLE 4-1 Heating application data ratings— Air temperature entering outdoor coil degree °F. Bard Manufacturing Company

OUTDOOR MODEL	INDOOR COIL		−5°	0°	5°	10°	15°	20°	25°	30°	35°	40°	45°	47°	50°	55°	60°	65°
24UHPQB	BC24B	Btuh	8,140	11,100	12,000	13,000	14,000	14,900	15,900	16,900	17,884	20,600	23,400	24,600	25,600	27,300	29,000	30,700
		Watts	2,025	2,069	2,114	2,158	2,200	2,246	2,290	2,334	2,379	2,423	2,467	2,485	2,511	2,555	2,599	2,644
		COP	1.17	1.57	1.66	1.76	1.86	1.94	2.03	2.12	2.20	2.49	2.77	2.90	2.98	3.13	3.26	3.40
	A36AQ-A	Btuh	7,361	9,414	11,000	11,800	12,805	14,166	15,527	16,888	18,250	20,479	22,708	23,600	24,540	26,100	27,673	29,240
		Watts	1,789	1,839	1,895	1,922	1,964	2,005	2,047	2,088	2,130	2,171	2,213	2,230	2,254	2,296	2,337	2,379
		COP	1.20	1.50	1.70	1.80	1.91	2.07	2.22	2.36	2.51	2.76	3.00	3.10	3.18	3.33	3.46	3.60
30UHPQB	BC36B	Btuh	7,000	8,700	11,500	12,400	14,300	16,100	17,900	19,800	21,700	24,700	27,700	29,000	30,400	32,700	35,000	37,400
		Watts	1,700	1,932	2,080	2,217	2,329	2,408	2,476	2,546	2,597	2,713	2,801	2,835	2,866	2,905	2,949	2,988
		COP	1.20	1.32	1.48	1.64	1.80	1.96	2.12	2.28	2.45	2.67	2.90	3.00	3.11	3.30	3.48	3.67
	A36A(Q,S)-A	Btuh	7,800	9,500	11,300	13,000	14,700	16,400	18,100	19,800	21,600	25,000	28,400	29,800	31,100	33,300	35,600	37,800
		Watts	1,662	1,820	1,948	2,129	2,256	2,357	2,446	2,524	2,606	2,756	2,871	2,913	2,942	2,986	3,034	3,070
		COP	1.37	1.53	1.70	1.79	1.91	2.04	2.17	2.30	2.43	2.66	2.90	3.00	3.10	3.27	3.44	3.61
	A37AQ-A	Btuh	7,700	8,783	10,900	13,000	15,100	17,250	19,300	21,400	23,600	25,800	28,100	29,000	30,200	32,200	34,200	36,200
		Watts	2,032	2,109	2,186	2,263	2,340	2,417	2,493	2,570	2,647	2,727	2,801	2,832	2,878	2,954	3,031	3,108
		COP	1.11	1.22	1.46	1.68	1.89	2.09	2.26	2.43	2.61	2.77	2.93	3.00	3.07	3.19	3.30	3.41
36UHPQB	BC36B	Btuh	10,500	12,400	14,300	16,200	18,000	19,900	21,800	23,700	25,600	28,900	32,900	34,000	35,400	37,700	40,000	42,400
		Watts	2,106	2,258	2,410	2,540	2,639	2,739	2,828	2,908	2,967	3,070	3,116	3,190	3,244	3,280	3,304	3,342
		COP	1.46	1.61	1.74	1.87	2.00	2.13	2.26	2.39	2.53	2.76	3.05	3.10	3.20	3.37	3.55	3.72
	A36AQ-B	Btuh	11,100	12,900	14,700	16,400	18,100	19,800	21,600	23,300	25,000	29,000	32,700	33,600	35,000	37,200	39,500	41,800
		Watts	2,017	2,186	2,355	2,504	2,627	2,751	2,866	2,970	3,054	3,309	3,424	3,397	3,444	3,496	3,553	3,605
		COP	1.61	1.73	1.83	1.92	2.02	2.11	2.21	2.30	2.40	2.57	2.80	2.90	2.98	3.12	3.26	3.40
	A37AQ-A	Btuh	11,722	13,544	15,366	17,188	19,011	20,833	22,655	24,477	26,300	30,341	34,383	36,000	37,500	40,000	42,500	45,000
		Watts	2,356	2,457	2,557	2,658	2,758	2,859	2,959	3,060	3,160	3,261	3,361	3,402	3,462	3,562	3,663	3,763
		COP	1.45	1.61	1.76	1.89	2.00	2.13	2.24	2.34	2.43	2.72	2.99	3.10	3.17	3.29	3.39	3.50
42UHPQA	BC48B	Btuh	12,000	14,500	17,000	19,200	21,600	24,000	26,400	28,800	31,200	34,450	37,700	39,000	40,500	43,000	45,500	48,000
		Watts	3,040	3,080	3,120	3,240	3,320	3,400	3,470	3,550	3,625	3,700	3,780	3,812	3,860	3,935	4,010	4,090
		COP	1.15	1.38	1.60	1.73	1.90	2.07	2.23	2.37	2.52	2.73	2.92	3.00	3.07	3.20	3.32	3.44
	A61AQ-A	Btuh	15,300	17,200	19,100	20,900	22,700	24,600	26,400	28,300	30,100	34,600	39,100	41,000	42,600	45,300	47,900	50,600
		Watts	3,170	3,210	3,250	3,280	3,310	3,330	3,380	3,420	3,450	3,490	3,520	3,536	3,550	3,590	3,625	3,660
		COP	1.41	1.57	1.72	1.86	2.01	2.15	2.29	2.42	2.55	2.90	3.26	3.40	3.50	3.70	3.87	3.90
48UHPQB	BC48B	Btuh	15,800	18,000	20,200	22,800	25,000	28,200	31,400	34,600	37,800	41,400	45,000	46,500	48,300	51,400	54,500	57,600
		Watts	3,600	3,650	3,700	3,900	4,065	4,160	4,260	4,360	4,460	4,560	4,660	4,700	4,760	4,860	4,960	5,060
		COP	1.28	1.45	1.60	1.71	1.80	1.98	2.16	2.32	2.48	2.66	2.83	2.90	2.97	3.10	3.22	3.33
48UHPQB-B	A61AQ-A	Btuh	16,600	19,710	22,680	34,390	26,300	28,700	31,200	33,600	36,100	41,000	46,000	48,000	49,900	53,000	56,200	59,400
		Watts	3,790	3,850	3,910	3,970	4,020	4,085	4,140	4,200	4,260	4,320	4,370	4,400	4,435	4,500	4,550	4,610
		COP	1.28	1.50	1.70	1.80	1.91	2.06	2.21	2.34	2.48	2.66	3.08	3.20	3.29	3.45	3.62	3.77
60UHPQB	BC60B	Btuh	18,500	21,700	24,900	28,000	31,200	34,400	37,600	40,700	43,900	49,800	55,600	58,000	60,300	64,200	68,100	72,100
		Watts	4,066	4,190	4,314	4,430	4,550	4,670	4,790	4,900	5,029	5,150	5,260	5,315	5,380	5,500	5,625	5,740
		COP	1.33	1.51	1.69	1.85	2.01	2.16	2.30	2.43	2.55	2.83	3.09	3.20	3.28	3.42	3.55	3.68
60UHPQB-B	A61AQ-A	BUTH	18,400	21,700	25,200	22,680	32,000	35,500	38,900	42,300	45,800	52,100	58,500	61,000	63,500	67,800	72,000	76,300
		Watts	4,100	4,245	4,355	3,910	4,670	4,810	4,960	5,100	5,250	5,390	5,530	5,590	5,675	5,810	5,960	6,100
		COP	1.31	1.49	1.70	1.70	2.00	2.16	2.29	2.43	2.55	2.83	3.10	3.20	3.28	3.42	3.54	3.66

70 degree F DB Return Air at Rated CFM.

For additional cooling and heating application data refer to form F1275.

Table 4-2 Cooling application data ratings. Bard Manufacturing Company

24UHPQB Outdoor Model

INDOOR MODEL	D.B. W.B	COOLING CAPACITY	75°	85°	95°	105°	115°
RC24R	80/67	Total Cooling	25,200	24,600	23,600	22,200	20,500
		Sensible Cooling	17,300	17,100	16,900	16,500	16,000
A36AQ-A	80/67	Total Cooling	24,700	24,300	23,000	21,800	20,300
		Sensible Cooling	18,000	17,600	17,250	17,000	16,700

(OUTDOOR TEMPERATURE °F)

36UHPQB Outdoor Model

INDOOR MODEL	D.B. W.B	COOLING CAPACITY	75°	85°	95°	105°	115°
BC36B	80/67	Total Cooling	39,380	36,690	34,000	31,310	28,620
		Sensible Cooling	26,790	25,950	25,100	24,250	23,410
A36AQ-B	80/67	Total Cooling	37,900	35,690	33,000	29,840	26,700
		Sensible Cooling	25,920	26,520	25,600	23,160	19,200
A37AQ-A	80/67	Total Cooling	42,200	39,100	36,000	32,900	29,800
		Sensible Cooling	27,900	26,900	25,920	24,900	24,000

(OUTDOOR TEMPERATURE °F)

48UHPQB Outdoor Model

INDOOR MODEL	D.B. W.B	COOLING CAPACITY	75°	85°	95°	105°	115°
BC48B	80/67	Total Cooling	49,800	47,900	45,500	42,600	39,300
		Sensible Cooling	34,000	33,700	33,200	32,600	31,800
A61AQ-A	80/67	Total Cooling	52,400	51,700	50,000	47,400	43,800
		Sensible Cooling	37,300	36,700	36,000	35,100	34,100

(OUTDOOR TEMPERATURE °F)

30UHPQB Outdoor Model

INDOOR MODEL	D.B. W.B	COOLING CAPACITY	75°	85°	95°	105°	115°
RC36R	80/67	Total Cooling	34,290	31,490	29,000	26,800	24,900
		Sensible Cooling	24,440	23,130	22,000	21,060	20,300
A36A(3,5)-A	80/67	Total Cooling	32,210	30,420	28,200	25,560	22,500
		Sensible Cooling	24,700	24,330	23,400	21,920	19,900
A37AQ-A	80/67	Total Cooling	34,500	32,300	30,000	27,600	25,000
		Sensible Cooling	23,300	22,600	21,900	21,300	20,700

(OUTDOOR TEMPERATURE °F)

42UHPQA Outdoor Model

INDOOR MODEL	D.B. W.B	COOLING CAPACITY	75°	85°	95°	105°	115°
BC48H	80/67	Total Cooling	42,800	42,200	40,500	37,700	33,800
		Sensible Cooling	31,600	31,000	30,300	29,300	27,700
A61AQ-A	80/67	Total Cooling	47,200	46,100	43,500	41,000	37,100
		Sensible Cooling	33,200	32,800	32,000	31,500	29,600

(OUTDOOR TEMPERATURE °F)

60UHPQB Outdoor Model

INDOOR MODEL	D.B. W.B	COOLING CAPACITY	75°	85°	95°	105°	115°
BC60B	80/67	Total Cooling	60,500	59,600	57,500	54,700	49,700
		Sensible Cooling	41,000	40,700	40,200	39,600	38,800
A61AQ-A	80/67	Total Cooling	61,700	60,500	58,000	54,200	49,200
		Sensible Cooling	41,200	40,700	40,000	39,000	37,800

(OUTDOOR TEMPERATURE °F)

Table 4-3 Specifications for a split heat pump. Bard Manufacturing Company

MODEL	24UHPQB	30UHPQB	36UHPQB	42UHPQB	48UHPQB	48UHPQB-B	60UHPQB	60UHPQB-B
Electrical rating (60 Hz/V/Ph)	230/208-1	230/208-1	230/208-1	230/208-1	230/208-1	230/208-1	230/208-1	230/208-1
Operating voltage range	197-253	197-253	197-253	197-253	197-253	197-253	197-253	197-253
Minimum circuit ampacity	18	21.5	24	26.4	36	21.4	41	27.2
*Field wire size	#12	#10	#10	#10	#8	#12	#6	#10
**Delay fuse max. or ckt. bkr.	30	30	40	45	60	35	60	45
Total unit amps 230/208	12.6/14.1	14.5/17.4	16.4/18.6	19.3/19.0	25.8/28.5	15.7/17.4	28.9/33.2	18.3/22.2
Compressor								
Volts	230/208	230/208	230/208	230/208	230/208	230/208	230/208	230/208
Rated load amps 230/208	11.5/13.0	13.4/16.3	15.3/17.5	17.8/17.5	24.3/27	14.2/15.9	26.6/30.9	16.0/19.9
Branch circuit selection current	13	14.8	18	19.9	27	15.9	30.9	19.9
Lock rotor amps 230/208	62.5/62.5	76/76	83.5/83.5	107/107	129/129	99/99	169/169	123/123
Crankcase heat	N/A	Solid state immersion	Solid state immersion	Solid state immersion	Solid state immersion	Solid state immersion	N/A	N/A
Fan motor & condenser								
Fan motor–HP/RPM	1/6 - 825	1/6 - 825	1/6 - 825	1/4 - 825	1/4 - 825	1/4 - 825	1/3 - 1075	1/3 - 1075
Fan motor–AMPS	1.1	1.1	1.1	1.5	1.5	1.5	2.3	2.3
Fan–DIA/CFM	24" - 3000	24" - 3000	24" - 3000	24" - 3100	24" - 3100	24" - 3100	24" - 3400	24" - 3400
Face area sq.ft./row	13.2/1	13.2/1	13.2/1	13.2/2	13.2/2	13.2/2	17.1/2	17.1/2
Fins per in.	14	14	13	13	13	13	13	13
Refrigerant connection & charge quick connect system								
Suction line fitting	–10	–10	–10	–12	–12	–12	–12	–12
Liquid line fitting	–6	–6	–6	–6	–6	–6	–6	–6
Factory charge (Oz.)	96	84	111	170	192	192	214	214
Shipping weight lbs.	200	200	200	230	245	245	305	305

*60 degree C Copper wire size.

**Maximum time delay fuse or HACR type circuit breaker

Table 4-4 Performance data. The Trane Company, an American Standard Company

Indoor Fan Performance WCC018F
External Static Pressure

AIRFLOW CFM [2]	HIGH SPEED			LOW SPEED [1]		
	PRESS IN W G	PWR. WATTS	BHP	PRESS IN. W G	PWR. WATTS	BHP
450	0.75	229	0.18	0.6	166	0.13
475	0.73	233	0.19	0.58	170	0.14
500	0.72	236	0.19	0.55	174	0.14
525	0.7	240	0.19	0.52	177	0.14
550	0.68	243	0.19	0.5	181	0.14
575	0.67	247	0.20	0.46	186	0.15
600	0.65	250	0.20	0.43	190	0.15
625	0.63	254	0.20	0.39	195	0.16
650	0.61	257	0.21	0.35	199	0.16
675	0.6	261	0.21	0.3	205	0.16
700	0.58	265	0.21	0.24	211	0.17
725	0.56	—	—	0.13	221	0.18
—	—	—	—	—	—	—
—	—	—	—	—	—	—
—	—	—	—	—	—	—
—	—	—	—	—	—	—

[1] FACTORY SETTING AT LOW SPEED
[2] WET COIL, NO FILTER, NO HEATER
SEE PRESSURE DROP TABLES FOR INSTALLED ELECTRIC HEATER.

From Dwg. 21A730144 Rev. 0

Indoor Fan Performance WCC024F
External Static Pressure

AIRFLOW CFM [3]	HIGH SPEED [1]			LOW SPEED		
	PRESS IN. W G.	PWR. WATTS	BHP	PRESS IN W G	PWR. WATTS	BHP
600	0.67	250	0.20	0.46	190	0.20
625	0.65	254	0.20	0.42	195	0.20
650	0.63	257	0.21	0.36	201	0.21
675	0.62	261	0.21	0.29	208	0.21
700	0.60	265	0.21	0.2	216	0.21
725	0.58	268	0.21	—	—	—
750	0.56	272	0.22	—	—	—
775	0.54	276	0.22	—	—	—
800	0.51	280	0.22	—	—	—
825	0.49	284	0.23	—	—	—
850	0.47	288	0.23	—	—	—
875	0.44	292	0.23	—	—	—
900 [2]	0.41	296	0.24	—	—	—
925	0.39	301	0.24	—	—	—
950	0.36	305	0.24	—	—	—
975	0.32	310	0.25	—	—	—
1000	0.29	315	0.25	—	—	—

[1] FACTORY SETTING AT HI SPEED
[2] WATER BLOW-OFF LIMIT
[3] WET COIL, NO FILTER, NO HEATER
SEE PRESSURE DROP TABLES FOR INSTALLED ELECTRIC HEATER

From Dwg. 21A730146 Rev. 1

Indoor Fan Performance WCC030F
External Static Pressure

AIRFLOW CFM [2]	HIGH SPEED			LOW SPEED [1]		
	PRESS. IN W G	PWR WATTS	BHP	PRESS IN. W G	PWR. WATTS	BHP
750	0.95	316	0.25	0.79	322	0.26
800	0.92	331	0.26	0.74	337	0.27
850	0.88	346	0.28	0.68	352	0.28
900	0.84	361	0.29	0.62	368	0.29
950	0.8	376	0.30	0.56	384	0.31
1000	0.76	391	0.31	0.49	400	0.32
1050	0.72	406	0.32	0.4	418	0.33
1100	0.68	421	0.34	0.29	437	0.35
1150	0.63	436	0.35	0.15	463	0.37
—	—	—	—	—	—	—
—	—	—	—	—	—	—
—	—	—	—	—	—	—
—	—	—	—	—	—	—
—	—	—	—	—	—	—
—	—	—	—	—	—	—
—	—	—	—	—	—	—

[1] FACTORY SETTING AT LOW SPEED
[2] WET COIL, NO FILTER, NO HEATER
SEE PRESSURE DROP TABLES FOR INSTALLED ELECTRIC HEATER

From Dwg 21A730240 Rev. 0

Indoor Fan Performance WCC036F
External Static Pressure

AIRFLOW CFM [3]	HIGH SPEED [1]			LOW SPEED		
	PRESS. IN. W G.	PWR. WATTS	BHP	PRESS IN W G	PWR WATTS	BHP
900	0.84	381	0.29	0.62	377	0.29
950	0.8	392	0.30	0.56	388	0.31
1000	0.76	404	0.31	0.49	400	0.32
1050	0.72	416	0.32	0.4	413	0.33
1100	0.68	427	0.34	0.29	427	0.35
1150	0.63	439	0.35	0.15	439	0.37
1200	0.58	450	0.36	—	—	—
1250	0.52	462	0.37	—	—	—
1300	0.46	473	0.38	—	—	—
1350 [2]	0.38	485	0.40	—	—	—
1400	0.3	498	0.41	—	—	—
1450	0.19	507	0.42	—	—	—
1500	0.07	518	0.43	—	—	—
—	—	—	—	—	—	—
—	—	—	—	—	—	—
—	—	—	—	—	—	—

[1] FACTORY SETTING AT HI SPEED
[2] WATER BLOW-OFF LIMIT
[3] WET COIL, NO FILTER, NO HEATER
SEE PRESSURE DROP TABLES FOR INSTALLED ELECTRIC HEATER

From Dwg. 21A730242 Rev. 0

Indoor Fan Performance WCC042F
External Static Pressure

AIRFLOW CFM³	HIGH SPEED¹			LOW SPEED		
	PRESS. IN. W.G.	PWR. WATTS	BHP	PRESS IN. W.G.	PWR. WATTS	BHP
1050	0.76	410	0.33	.45	410	.33
1100	0.73	422	0.34	.39	422	.34
1150	0.69	435	0.35	.33	435	.35
1200	0.65	448	0.36	.27	449	.36
1250	0.60	462	0.37	.19	459	.37
1300	0.54	477	0.38	.11	470	.38
1350	0.49	493	0.39	—	—	—
1400	0.42	510	0.41	—	—	—
1450	0.35	528	0.42	—	—	—
1500²	0.27	550	0.44	—	—	—
1550	0.18	571	0.46	—	—	—
1600	0.08	593	0.47	—	—	—
—	—	—	—	—	—	—
—	—	—	—	—	—	—
—	—	—	—	—	—	—
—	—	—	—	—	—	—

¹ FACTORY SETTING AT HI SPEED
² WATER BLOW-OFF LIMIT
³ WET COIL, NO FILTER, NO HEATER.
SEE PRESSURE DROP TABLES FOR INSTALLED ELECTRIC HEATER.

From Dwg. 21A730858 Rev. 1

Indoor Fan Performance WCC048F
External Static Pressure

AIRFLOW CFM²	HIGH SPEED			LOW SPEED¹		
	PRESS. IN. W.G.	PWR. WATTS	BHP	PRESS IN. W.G.	PWR. WATTS	BHP
1200	—	—	—	1.00	616	.465
1250	—	—	—	0.96	630	.473
1300	—	—	—	0.92	644	.481
1350	—	—	—	0.87	658	.487
1400	1.05	745	.568	0.82	673	.493
1450	1.01	761	.588	0.77	688	.497
1500	0.97	777	.608	0.71	703	.499
1550	0.94	793	.628	0.65	720	.500
1600	0.90	810	.648	0.58	737	.498
1650	0.85	826	.667	0.51	755	.492
1700	0.81	843	.686	0.41	775	.480
1750	0.77	861	.704	0.28	800	.454
1800	0.72	878	.722	—	—	—
—	—	—	—	—	—	—
—	—	—	—	—	—	—
—	—	—	—	—	—	—

¹ FACTORY SETTING AT LOW SPEED
² WET COIL, NO FILTER, NO HEATER.
SEE PRESSURE DROP TABLES FOR INSTALLED ELECTRIC HEATER.

From Dwg. 21A730902 Rev. 0

Indoor Fan Performance WCC060F
External Static Pressure

AIRFLOW CFM²	HIGH SPEED			LOW SPEED¹		
	PRESS. IN. W.G.	PWR. WATTS	BHP	PRESS IN. W.G.	PWR. WATTS	BHP
1400	1.06	745	.568	0.86	715	.499
1450	1.02	761	.588	0.79	731	.497
1500	0.98	777	.608	0.73	747	.493
1550	0.95	793	.628	0.67	763	.488
1600	0.91	810	.648	0.61	779	.482
1650	0.86	826	.667	0.56	795	.475
1700	0.82	843	.686	0.50	811	.467
1750	0.78	861	.704	0.45	827	.458
1800	0.73	878	.722	0.40	843	.448
1850	0.69	896	.740	0.35	859	.438
1900	0.64	915	.757	0.30	875	.427
1950	0.58	934	.773	0.24	891	.416
2000	0.53	953	.789	0.20	907	.405
2050	0.47	973	.804	—	—	—
2100	0.41	994	.818	—	—	—
2150	0.35	1016	.831	—	—	—
2200²	0.28	1038	.842	—	—	—
—	—	—	—	—	—	—

¹ FACTORY SETTING AT HIGH SPEED
² WATER BLOW-OFF LIMIT
³ WET COIL, NO FILTER, NO HEATER.
SEE PRESSURE DROP TABLES FOR INSTALLED ELECTRIC HEATER.

From Dwg. 21A730903 Rev. 0

Indoor Fan Performance
Hi-Static Motor Accessory — BAYHSMT043A
WCC030F, WCC036F
External Static Pressure

AIRFLOW CFM²	230V/460V			208V		
	PRESS. IN. W.G.	PWR. WATTS	BHP	PRESS IN. W.G.	PWR. WATTS	BHP
900	1.76	724	0.57	1.70	671	0.53
950	1.68	744	0.59	1.57	689	0.54
1000	1.61	764	0.60	1.45	708	0.56
1050	1.53	785	0.62	1.31	726	0.58
1100	1.45	805	0.64	1.17	745	0.59
1150	1.36	826	0.66	1.03	764	0.60
1200	1.27	846	0.67	0.88	783	0.62
1250	1.17	866	0.69	0.72	803	0.63
1300	1.06	887	0.70	0.55	822	0.65
1350¹	0.95	907	0.72	—	—	—
1400	0.82	928	0.73	—	—	—
1450	0.68	949	0.75	—	—	—
1500	0.50	969	0.77	—	—	—
—	—	—	—	—	—	—
—	—	—	—	—	—	—
—	—	—	—	—	—	—

¹ WATER BLOW-OFF LIMIT
² WET COIL, NO FILTER, NO HEATER.

From Dwg. 21A730914 Rev. 0

Table 4-5 General data for a split system.
The Trane Company, an American-Standard Company

OUTDOOR UNIT	TWX024C100A
POWER CONNS. — V/Ph/Hz	200/230/1/60
Min. Brch. Cir. Ampacity③	15
Br. Cir. ⎰ Max. (Amps)	25
Prot. Rtg. ⎱ Recmd. (Amps)	20
NOISE RATING NO.②	7.8
COMPRESSOR	CLIMATUFF™
No. Used — No. Speeds	1 — 1
Volts/Ph/Hz	200/230/1/60
R.L. Amps — L.R. Amps	9 — 57
Brch. Cir. Select. Cur. Amps	11.0
OUTDOOR FAN — Type	PROPELLER
Dia. (In.) — No. Used	22 — 1
Type Drive — No. Speeds	DIRECT — 2
CFM @ 0.0 in. w.g.④	1990
No. Motors — HP	1 — 1/6
Motor Speed R.P.M.	825
Volts/Ph/Hz	200/230/1/60
F.L. Amps	0.90
OUTDOOR COIL — Type	SPINE FIN™
Rows — F.P.I.	1 — 24
Face Area (Sq. Ft.)	17.53
Tube Size (in.)	3/8
Refrigerant Control	EXPANSION VALVE
REFRIGERANT	
Lbs. — R-22 (O.D. Unit)⑤	8-LBS.,11-OZ.
Factory Supplied	YES
Line Size — in. O.D. Gas⑥	3/4
iLine Size — in. O.D. Liq.⑥	5/16
DIMENSIONS	H X W X D
Outdoor Unit — Crated (in.)	37-1/2 X 35 X 31
Uncrated	SEE OUTLINE DWG.
WEIGHT Shipping (lbs.) — Net (lbs.)	268 - 240

① Rated in accordance with A.R.I. Standard 240.

② Rated in accordance with A.R.I. Standard 270.

SPLIT SYSTEM

③ Calculated in accordance with National Electric Code. Suitable for use with HACR circuit breakers or fuses.

④ Standard Air — Dry Coil — Outdoor.

⑤ This value approximate. For more precise value see unit nameplate and service instruction.

⑥ Max. linear length 80 ft.; Max. lift — Suction 60 ft.; Max. lift — Liquid 60 ft. Max. length of precharged tubing 50 ft. For greater length refer to Refrigerant Piping Manual Pub. No. 32-3009.

⑦ Rated in Accordance with D.O.E. test procedure.

OUTDOOR UNIT WITH HEAT PUMP COILS

	TXM725B4	TXM736B4	TXS024A4	TXS724A4
EXPANSION TYPE	CHG TO 61	CHG TO 61	CHG TO 61	CHG TO 61
RATINGS (Cooling)①				
BTUH (Total)	22600	23400	22600	22600
BTUH (Sensible)	14100	14600	14100	14100
Indoor Airflow (CFM)	800	885	800	800
System Power (KW)	2.24	2.28	2.24	2.24
SEER (BTU/Watt-Hr.)⑦	10.80	10.90	10.75	10.70
RATINGS (Heating)①				
(High Temp.) BTUH	20800	21400	21000	21000
System Power (KW)	2.07	2.06	2.07	2.07
COP	2.94	3.04	2.98	2.38
HSPF (BTU/Watt-Hr.)⑦	7.00	7.15	7.05	7.05

OUTDOOR UNIT WITH AIR HANDLERS

	TWE036C14	TWH024B14	TWH030B14	TWH036B14	TWH039E15-C	TWH039P15-C	TWH724B14	TWH730B14
EXPANSION TYPE	CHG TO 61	CHG TO 61	CHG TO 61	CHG TO 61	TXV-B	TXV-B	CHG TO 61	CHG TO 61
RATINGS (Cooling)①								
BTUH (Total)	23800	23600	24400	24600	26200	25800	22400	23200
BTUH (Sensible)	14900	14700	15200	15400	16400	16100	14000	14500
Indoor Airflow (CFM)	875	800	895	900	850	900	800	850
System Power (KW)	2.32	2.23	2.32	2.33	2.14	2.30	2.19	2.25
SEER (BTU/Watt-Hr.)⑦	11.15	11.55	11.50	11.50	13.60	12.50	11.10	11.20
RATINGS (Heating)①								
(High Temp.) BTUH	21400	21000	21600	21600	21400	22000	21000	21400
System Power (KW)	2.10	2.04	2.06	2.03	1.78	1.93	2.04	2.04
COP	2.98	3.02	3.08	3.12	3.52	3.34	3.02	3.08
HSPF (BTU/Watt-Hr.)⑦	7.05	7.15	7.20	7.30	8.00	7.65	7.15	7.25

+See page 38 and 39 for combinations with Auxiliary Devices.

OUTDOOR UNIT	TWX024C100A
POWER CONNS. — V/Ph/Hz	200/230/1/60
Min. Brch. Cir. Ampacity③	15
Br. Cir. } Max. (Amps)	25
Prot. Rtg. } Recmd. (Amps)	20
NOISE RATING NO.②	7.8
COMPRESSOR	CLIMATUFF™
No. Used — No Speeds	1 — 1
Volts/Ph/Hz	200/230/1/60
R.L. Amps — L.R. Amps	9 — 57
Brch. Cir. Select. Cur. Amps	·· ?
OUTDOOR FAN — Type	PROPELLER
Dia. (In.) — No. Used	22 — 1
Type Drive — No. Speeds	DIRECT — 2
CFM @ 0.0 in. w.g.④	1990
No. Motors — HP	1 — 1/6
Motor Speed R.P.M.	825
Volts/Ph/Hz	200/230/1/60
F.L. Amps	0.90
OUTDOOR COIL — Type	SPINE FIN™
Rows — F.P.I.	1 — 24
Face Area (Sq. Ft.)	17.53
Tube Size (in.)	3/8
Refrigerant Control	EXPANSION VALVE
REFRIGERANT	
Lbs. — R-22 (O.D. Unit)⑤	8-LBS.,11-OZ.
Factory Supplied	YES
Line Size — in. O.D. Gas⑥	3/4
Line Size — in. O.D. Liq. ⑥	5/16
DIMENSIONS	H X W X D
Outdoor Unit — Crated (in.)	37-1/2 X 35 X 31
Uncrated	SEE OUTLINE DWG.
WEIGHT Shipping (lbs.) — Net (lbs.)	268 - 240

① Rated in accordance with A.R.I. Standard 240.

② Rated in accordance with A.R.I. Standard 270.

SPLIT SYSTEM

③ Calculated in accordance with National Electric Code. Suitable for use with HACR circuit breakers or fuses.

④ Standard Air — Dry Coil — Outdoor.

⑤ This value approximate. For more precise value see unit nameplate and service instruction.

⑥ Max. linear length 80 ft.; Max. lift — Suction 60 ft.; Max. lift — Liquid 60 ft. Max. length of precharged tubing 50 ft. For greater length refer to Refrigerant Piping Manual Pub. No. 32-3009.

⑦ Rated in Accordance with D.O.E. test procedure.

OUTDOOR UNIT WITH AIR HANDLERS

	TWH739E15-C	TWH739P15-C	TWV024B14	TWV025B14	TWV030B14	TWV036B14	TWV039E15-C	TWV039P15-C
EXPANSION TYPE	TXV-B	TXV-B	CHG TO 61	CHG TO 61	CHG TO 61	CHG TO 61	TWV-B	TXV-B
RATINGS (Cooling)①								
BTUH (Total)	26200	26000	23400	23400	24400	24600	25600	26000
BTUH (Sensible)	16400	16200	14600	14600	15200	15400	16000	16200
Indoor Airflow (CFM)	850	900	800	800	895	900	800	900
System Power (KW)	2.15	2.27	2.28	2.24	2.31	2.39	2.18	2.32
SEER (BTU/Watt-Hr.)⑦	13.55	12.70	11.20	11.50	11.50	11.20	13.25	12.50
RATINGS (Heating)①								
(High Temp.) BTUH	21400	21800	21000	21000	21600	21800	21600	22000
System Power (KW)	1.78	1.89	2.07	2.05	2.06	2.07	1.78	1.93
COP	3.52	3.38	2.98	3.00	3.08	3.08	3.56	3.34
HSPF (BTU/Watt-Hr.)⑦	8.00	7.75	7.05	7.10	7.25	7.15	7.60	7.65

	TWV724B14	TWV725B14	TWV730B14	TWV739E15-C	TWV739P15-C
EXPANSION TYPE	CHG TO 61	CHG TO 61	CHG TO 61	TWV-B	TXV-B
RATINGS (Cooling)①					
BTUH (Total)	22200	22400	23200	25600	25800
BTUH (Sensible)	13900	14000	14500	16000	16100
Indoor Airflow (CFM)	800	800	850	850	900
System Power (KW)	2.24	2.22	2.27	2.12	2.35
SEER (BTU/Watt-Hr.)⑦	10.75	11.00	11.15	13.25	12.15
RATINGS (Heating)①					
(High Temp.) BTUH	21000	21000	21400	21600	22000
System Power (KW)	2.07	2.05	2.04	1.78	1.95
COP	2.98	3.00	3.08	3.56	3.30
HSPF (BTU/Watt-Hr.)⑦	7.05	7.10	7.20	7.80	7.55

†See pages 36 and 37 for combinations with Auxiliary Devices.

Table 4-6 Performance data for heating and cooling.
The Trane Company, an American-Standard Company

Performance data* cooling	Evaporating air temp, °F.		Operating pressures		Electrical ratings		R-22 refrig.	Comp. oil
	Discharge air	Temp drop °F.	Suction	Discharge	Amps	Locked rotor amps	Charge in ounces	Charge in fluid oz.
ES11H33A	56.6	23.4	76	270	5.1 5.5	26.0	20	9.13
ES13H33A	53.7	26.3	75	295	6.2 6.6	31.0	28	10.82
ES15H33A	51.0	29.0	72	295	7.5 8.2	42.0	30	13.9
EL19H35B	57.6	22.4	76	276	8.8 9.5	42.0	38	14.5
EL24H35	54.1	25.9	77	282	11.7 12.9	104.0	47	34.0
EL31H35B	51.09	28.91	73	310	16.9 18.3	83.5	78	55.0

*Rating conditions: 80°F. room temperature and 50% relative humidity with
90°F. outside temperature at 40% relative humidity.

Performance data heating	Volts	Btuh	CFM high speed	Heat rise
ES11H33A	230 208	10700 8900	300	33.0
ES13H33A	230 208	10700 8900	320	31.0
ES15H33A	230 208	10700 8900	320	31.0
EL19H35B	230 208	17300 14300	570	28.1
EL24H35	230 208	17300 14300	590	27.5
EL31H35B	230 208	17300 14300	625	25.6

Table 4-7 Common types of heat pumps

Air-to-air
Air-to-water
Water-to-water
Water-to-air

Table 4-8 Basic steps in choosing a heat pump

Have a load calculation done for cooling and heating needs
Shop for a desirable brand and type of heat pump
Decide on air terminal requirements
Evaluate outdoor air requirements
Choose duct locations and have ductwork sized

Choose locations for indoor and outdoor equipment
Confirm that installation requirements can be met
Choose equipment controls
Evaluate initial cost of the system
Evaluate efficiency and operating costs for the system

Table 4-9 Profile of an air-source heat pump

	Good
Stability	Extreme
Availability	Excellent
Initial cost	Low
Operating cost	High
Drawbacks	Frosting

Table 4-10 Profile of an earth-source heat pump

Suitability	Good
Stability	Stable
Availability	Excellent
Initial cost	Mid-range
Operating cost	Low to medium
Drawbacks	Leaks are hard to find and expensive to repair

Table 4-11 Profile of a surface-water-source heat pump

Suitability	Varies
Stability	Fair
Availability	Limited
Initial cost	Mid-range
Operating cost	Low
Drawbacks	Corrosion and dry spells

Table 4-12 Profile of a well-water-source heat pump

Suitability	Excellent
Stability	Stable
Availability	Very good
Initial cost	Mid-range
Operating cost	Low
Drawbacks	Mineral build-ups

Table 4-13 Profile of a solar-source heat pump

Suitability	Good
Stability	Extreme

Table 4-13 Continued

Availability	Excellent
Initial cost	Mid-range to high
Operating cost	Low
Drawbacks	Complicated system with expensive set-up costs

Table 4-14 Advantages of water-source heat pumps

No heating costs
Simplicity in design
Low maintenance costs
Longevity

Table 4-15 Disadvantages of water-source heat pumps

Potential for well or surface water drying up
Mineral build-ups
Corrosion
Possible water pump failure

Table 4-16 Types of solar-powered heat-pump systems

Solar-powered heat pumps
Passive solar heat pumps
Solar-assisted heat pumps
Solar evaporator coils

Table 4-17 Advantages of solar evaporator coils

Provides a high source of evaporator heat
Provides a high COP from the heat pump
Minimum auxiliary heat required

Table 4-18 Disadvantages of solar evaporator coils

Requires sunlight for best results
Minimum auxiliary heat required

Table 4-19 Advantages of passive solar systems

No moving parts to break or wear out
Low cost to operate
Few defrost cycles

Increases heat pump COP
No wind chill factor
Minimum auxiliary heat required

Table 4-20 Disadvantages of passive-solar systems

Space requirements can present problems
No other prominent disadvantages

Table 4-21 Advantages of solar-assisted heat pumps

Low operating costs
Provides a high COP from the heat pump
When coupled with storage facilities, it can operate
at offpeak electric rates

Table 4-22 Disadvantages of solar-assisted heat pumps

High initial costs
Complex design
Not efficient on cloudy days
Potential for high maintenance costs

5

Tools to have on hand

If you are planning to install your own heat pump, you need to know what tools to have on hand. The tools needed for installing a heat-pump system during new construction are about the same as those needed for heat-pump conversions. In both cases, there are some tools that you probably already have, and there are others that you might need to rent or buy.

Having the right tools for the job will not only make the work go smoother and faster, you will be adding to your personal safety. Many job-related accidents happen when mechanics use the wrong tool for the job. For instance, a heating mechanic who uses a screwdriver as a chisel might wind up with broken pieces of the screwdriver in an eye. The job should be done with a chisel and eye protection. Duct work can be very sharp, and using the wrong tools to cut it can result in an unwanted cut on the mechanic's hands or arms. There are many situations where using improper tools is not only inefficient, it is dangerous.

As a homeowner, it is unlikely that you will have a pair of power shears for cutting your duct work. You might have a reciprocating saw for cutting out holes to install floor registers. A homeowner might have one pair of metal-cutting snips, but professional tin benders use three pairs, all for different purposes. If you were to go out and buy all the specialty tools often used during heat-pump installations, you could spend hundreds of dollars. There is an alternative to buying expensive tools, you can rent them.

Renting tools

Renting tools is a wise way to get what you need when you only need it for a short period of time. Whether you need a power-driven tamping machine to prepare the ground for the heat-pump pad or a pair of power snips, you can rent the tools for a whole day at a fraction of their purchase price. Because you are not likely to be installing many heat pumps in the future, it makes good sense to rent these tools. All major tool-rental centers carry a large inventory of specialty tools. If you need it, they probably have it.

When you rent a tool, make sure the individual you rent it from explains how to operate it safely. A lot of times the employees of rental centers don't go into extreme detail on the operation of tools unless they are asked to. It is important that you become familiar with the tool before you take it to the job and use it.

Basic hand tools

There is a variety of hand tools that you should have available before beginning your installation or conversion work. Many of these tools are common and may already be in your tool box.

Hammer A hammer is needed for several parts of the job. Any type of carpenter's hammer will work. I prefer a heavy, framing hammer with a straight claw, but a lightweight hammer or a hammer with a curved claw is fine.

Tape measure A small, 12-foot tape will do, but a wide-blade, 25-foot tape measure is better. The larger tape allows you to make longer measurements, but most importantly, it can be extended for longer distances, unsupported, than a narrow-blade model can.

Pencils Pencils might not seem like much of a tool, but they will be needed throughout the course of your work. A flat, carpenter's pencil works best.

Nail pullers Nail pullers are often needed during the installation of duct work. If you are cutting out subflooring or through a floor joist, there might be nails in the path of your saw blade. I've found that small, rounded nail pullers, often called Cat's Paws, work best.

Tubing cutters A set of tubing cutters is needed when the time comes to work with the refrigeration tubing. These cutters have two roller wheels and a cutting wheel. They allow you to cut copper tubing so that the ends are cut squarely and smoothly.

Tubing cutters are available in various price ranges. Because you will not be using them often, you can get by with an inexpensive pair that can be purchased at most any hardware store.

Knife A sharp knife will be needed from time to time. One use for the knife is splitting foam insulation to be applied to refrigeration tubing, but you will discover numerous uses for the blade. A small pocket knife is all you will need.

A flashlight A flashlight might be needed with new work or conversion work. Conversion jobs often involve working in dark areas, and crawlspaces under new homes are difficult to see well in.

Electric droplight An electric droplight is very handy when working in dry crawlspaces. If there are puddles of water in the work area, resort to using a flashlight, to avoid any risk of electrocution.

Extension cord A good extension cord will be needed to run your power tools. You might need two cords, one for lighting and one to power the tools. A 50-foot cord is usually long enough, without being so bulky as to be a bother. Make sure the insulation on the cord is in good repair before using it.

Levels Levels are necessary for both the duct work and the setting of equipment. A 2-foot level and a torpedo level should see you through the job well. A torpedo level is a small, pocket-sized level that can be used when space will not allow the operation of a large level.

Screwdrivers You will need an assortment of screwdrivers. Some should be short and some should be long. It will also be necessary to have screwdrivers with both flat bits and Phillips-style bits. I've found that screwdrivers with rubber handles are the easiest to work with.

Tin snips It would be very difficult to install duct work without tin snips. The snips you use should be of a good quality, and they should have keen edges. When you select your snips, you will need three types. One pair of snips should be designed to cut straight lines. A second set of snips should be made to cut to the left, and the third set should be able to cut to the right. Don't try to skimp on your snips. You need all three types, and the better they are, the easier the job will be for you.

Yardstick A common yardstick can be used as a straightedge. If you don't have a yardstick, you can use a piece of lumber or anything else that is long enough and straight enough to allow you to mark even lines.

Scratch awl A scratch awl will be needed to mark your metal. Be careful when using the awl not to poke holes in yourself.

Drill bits Assorted drill bits for both metal and wood will be needed. Typically, an inexpensive set of bits from a hardware store will be all that is needed to give you the diversity in bit sizes and the cutting power needed.

A hole saw for your electric drill will be needed to cut a hole in the siding of the home. This hole will be needed to allow the refrigeration tubing and electrical wiring to pass through the outside wall and connect the inside unit with the outside unit.

Adjustable wrenches Adjustable wrenches are among the tools needed for HVAC work. A 12-inch wrench and a 6-inch wrench make a good combination.

Pliers When it comes to pliers, a couple of 12-inch tongue-and-groove pliers will come in handy. Some people call these pliers water-pump pliers. The jaws of the pliers angle out and away from the handles to accommodate many work scenarios.

Allen wrenches Allen wrenches should be in your toolbox before you set to work on your heat pump. An inexpensive assortment of sizes will be all that is needed.

Voltohmmeter A voltohmmeter will be needed before the job is done. These meters can be purchased at relatively low costs from many electronic stores, building centers, and hardware stores.

Fish tape A fish tape might be needed to fish wire through walls. If you are doing new-construction work, the fish tape won't be needed, but conversion work will require that you have some method of getting thermostat wire up through a wall.

Power tools

Power tools needed for installing duct work and heat pumps are not numerous, but they are important. The most crucial tool in this category is a reciprocating saw.

Reciprocating saws Reciprocating saws should be considered as necessities when installing heat pumps. These saws allow the user to cut holes quickly and with limited space. Cutting holes for all the floor registers in a home with a hand saw would seem to take forever. With the aid of a reciprocating saw, the same job can be done in a fraction of the time. Notching floor joists, cutting through sole plates and other necessary work is a breeze with the right saw, and a reciprocating saw is the right saw. Unfortunately, reciprocating saws are not cheap; they cost well over $100. These tools can, however, be rented at very reasonable rates.

In addition to the saw, you will need saw blades. There is a vast variety of blades available. They come in many lengths, and there are blades for cutting wood, plaster, metal, and more. Select your blades wisely and invest in the best.

Electric drills Electric drills are also needed when installing heat pumps. The drill doesn't have to be anything fancy. A ¼-inch drill is all that is needed. I prefer a ⅜-inch drill, but either one will work just fine.

Electric screw guns Electric screw guns are not a necessity, but they sure are a convenience. There are a lot of screws that go into an HVAC system, and turning them with electricity is much easier, and faster, than turning them by hand. As an alternative to buying a screw gun, you can simply buy a screw bit and chuck it into your electric drill.

Power shears Power shears, like electric screw guns, are not needed, but they are desirable when working with sheet metal. Electric shears are not likely to be on your list of want-to-buy tools, but they can be rented by the day at tool rental centers.

Other items

We have covered most of the mandatory tools, but there are some other items you should have on hand. One of which is a step ladder. Very few HVAC systems can be installed without working from a ladder at some point. Choose a ladder that meets safety requirements and don't climb above the recommended highest step. In addition to a ladder, there are a few other items you might need, especially if you are doing a conversion project.

Buckets Empty five-gallon buckets come in handy for a lot of things. They can be used to transport tools, to collect construction debris, or even as an improvised seat, and a lot more. I never leave home for a job without at least two buckets on my truck.

Tarps Tarps or rolls of plastic are very helpful during conversion projects. They can be used to protect floor covering from messy boots or to block dust out of rooms. They also make a good surface to lay tools and parts out on so that they won't get lost. If you are working in a crawl space, the tarp or plastic can be used as a ground cover.

Safety goggles Safety goggles or glasses should be considered mandatory equipment. With all the cutting and overhead work involved with installing a heat pump, a person's eyes are frequently at risk. For such a cheap investment, safety glasses provide immeasurable protection. Personally, I've found that safety glasses are more comfortable than goggles, and they don't steam up as badly.

Knee pads Knee pads provide protection to your knees when you are forced to work close to the floor. There isn't a lot of kneeling involved with a typical HVAC installation, but there is enough to justify the use of knee pads.

Gloves Thick work gloves can save you from a lot of cuts and scratches when working with sheet metal. Gloves might impede your ability to work to some degree, but they do offer protection to your hands. If you decide to work without gloves, buy a big box of adhesive bandages.

Duct tape Duct tape is used for so many purposes that people often forget that it is used with duct work. You will find duct tape to be very helpful in many parts of your job. For example, it can be used to tape the seams of tubing insulation together or to tape thermostat wire to your fish tape.

Pipe clamps You will need pipe clamps to secure the refrigeration tubing to the side of your home and to floor joists. There are several types of clamps to choose from.

Cable ties Cable ties are often used to attach wires to the refrigeration tubing. Duct tape can be used in lieu of cable ties, but the ties are less likely to come loose in the future.

Perforated strap iron Perforated strap iron works well as a temporary support for duct work. If you are working alone, the strap iron can act as a second set of hands. The material is sold in rolls and can be cut with tin snips.

Screws You are going to need a lot of self-taping sheet-metal screws. Make sure you buy screws that are self-taping; otherwise, you will be drilling numerous holes before you can install the screws.

6

Construction considerations for heat-pump conversions

The construction considerations for heat-pump conversions can present some problems. Anytime you must add or alter duct work in an existing home, the possibility for trouble exists. This is not to say that the job cannot be done. The question becomes one of what is the easiest, most effective, and least expensive way to provide suitable duct work for a new heat pump. To answer such a question, one must either have knowledge and experience with the task at hand, or they must consult experts that do.

Remodeling and conversion work is not a job that just anyone can do well. Professionals specialize in such fields, and it takes them years to gain the proficiency of an expert. Tradespeople who have worked new construction sometimes assume, wrongly, that if they can install systems in new homes, they can certainly upgrade old systems and perform complete conversion projects. There is a world of difference between working under the conditions new construction offers and those presented in the remodeling arena.

Homeowners usually have little or no experience with large-scale remodeling. Many homeowners have at least one or two areas of light remodeling that they can excel at. This, however, does not prepare them for the job of installing a complete heating and cooling system.

If you pick up your local phone directory and check the advertisement pages, you should find a significant number of heating and cooling contractors. Only a small percentage of those contractors will have the experience needed to provide a first-class conversion from ductless heat to ducted heat in a multi-story home.

Probably any of the contractors could handle the work involved in converting a one-story home from electric heat to a heat pump, but when you factor in the need to install duct work for additional stories of living space, the puzzle becomes much harder to solve.

If the home to be converted is receiving a complete interior re-modeling, meaning that the interior walls and ceilings will be re-worked, the job of adding or changing duct work is not difficult. But, when the home is to be left intact, finding ways to accomplish the duct work requirements without damaging exiting walls and ceilings is quite another question to ponder.

Is it possible to perform a complete conversion on a multistory home without any damage to existing walls and ceilings? It is unlikely that the most efficient heating system can be installed without some alterations to existing conditions, but skillful hands and minds can come up with ways to minimize damage.

Can a handy homeowner accomplish the task of installing new duct work in an old home? Yes; it is feasible that a homeowner can get the job done. The homeowner will usually need help, both from a knowledge point of view and from a second set of hands point of view, but the responsibility of installing the duct work with minimal disruption to the existing home is within the grasp of many home-owners.

This chapter cannot snip the tin for you or drive the connector strips, but it can give you the knowledge you need to progress with the conversion of your home from one type of heat to a heat pump. Whether you have a home equipped with old radiators, electric base-board heat, hot-water baseboard heat, or forced hot-air heat, this chapter shows you what to look out for in your conversion project.

Radiators

Radiators can be found in many old homes, but they have not been used in modern construction for a number of years. If your house is equipped with radiators, there is probably very little insulation in the home, and there is certainly no duct work. Both of these factors are something you will have to overcome in order to have a good heat-pump system. The windows and doors are other considerations.

Old houses were not built with energy-efficient windows and doors. Unless these components of your home have been upgraded somewhere along the line, you should give serious consideration to replacing them with new, insulated windows and doors. At the very least, you should install storm windows and doors.

Heat pumps work best when the demand on them for heat is low. The more you do to insulate the home, the better your new heat pump will work. In houses with radiators, this can mean substantial work and a pretty fair amount of cash.

A home that is heated with radiators will have a boiler some where, probably in the basement. To convert to a heat pump, you should remove the old radiators and boiler. The piping in the walls that feeds the radiators can remain. This is not a difficult job, but it is heavy work.

Homeowners can remove their own radiators, but due to the heavy heating units and the boiler, it is often wise to engage professionals to do the job. The water to the boiler must be turned off, and the entire heating system must be drained. This is done by opening drain valves at the boiler and cracking union nuts at the radiators to allow air into the system so that it will drain properly.

Once the system is drained, disconnecting the radiators is relatively easy. A pipe wrench is the only tool needed to get the radiators loose from the feed and return piping. If you do this part of the job yourself, be advised that the radiators are likely to be holding some dirty water. The black or rust-colored water will make quite a mess on a carpeted floor.

To avoid spilling ugly water on the floor, you should sit a pan under the connections as they are broken and then plug the openings in the radiators with screw-in plugs or caps. Old rags will do a reasonably good job as long as you don't expect them to hold the water in for too long.

The most difficult part of radiator removal is the sheer weight of the units. A strong appliance dolly can be used to move the radiators with minimal risk of injury. People with strong hands and backs can carry most residential radiators, but there is risk of personal injury, especially if stairs must be negotiated.

Radiators can usually be dismantled into sections to make their transportation less cumbersome. However, old radiators are often hard to get apart, especially if they have been painted. If the radiators are taken apart, there is a good chance they will leak when put back together. Because there is a good resale market for used radiators, most contractors take them out as complete units and sell them.

Once the radiators are gone, there are still pipes protruding from the floor. If a fitting can be seen close to the floor, the protruding pipes can be unscrewed and the fittings can be plugged with screw-in plugs. When the pipes run through the walls with no apparent fittings, the easiest method for removal is to cut them off. This is done best with a reciprocating saw that is equipped with a metal-cutting blade.

Many of the boilers used with radiators are huge. For this reason, the boilers are often disconnected and left in place. If you want to have the boiler removed, you should contract professionals to do the job.

Once the radiators are removed and the boiler is disconnected, you are ready to run duct work. Installing ducts for the first-floor living space should not be difficult. You will have access with either a basement, cellar, or crawl space to hang the duct work to the first-floor joists. It is getting the ducts to the second story that will require some creative installation methods.

There are many ways to route ducts to the upstairs. A common practice is to run the ducts up through a closet (or closets) on the first floor. Closets are natural duct chases that can be used without adversely affecting the living space of the home.

If you are unwilling to forfeit your closet space, you can run the ducts up and build a chase around them. This only involves finding a suitable location (usually a corner), extending the duct work up through the first floor, and framing an enclosure around it. Then drywall can be installed on the framing to make a finished wall.

Another creative way to get ducts upstairs is with the use of an outside chase. In this case, the duct work is run up the outside wall of the home. The chase is built around the duct work to imitate a chimney. From outward appearances, your house will seem to be equipped with a fireplace, but in reality, the chimney is just a hiding place for the duct work.

When the duct work has been routed to the upstairs, you still have some creative work to do. Getting it through the first level of living space unnoticed is only part of the job. Now you have to distribute the ducts throughout the upstairs with minimal damage.

The most simple way to install duct work in an existing second floor involves putting the ducts in the attic. This is the easiest way, but not the most efficient. Because heat rises, it is best to have the registers for the duct work in a floor. Putting them in a ceiling will work (many commercial buildings are done this way), but a floor-mounted register is better.

The advantage to putting the heat registers in the ceiling is the ease of installation. Once the chases have been made to get the ducts to the attic, the rest of the work is easy. All you have to do is lay the duct work on the ceiling joists and cut into the ceilings for the registers. Of course, the duct work installed outside of a home or in an attic should be wrapped with an insulation blanket.

Assuming that you don't want ceiling registers, the job gets much more complicated, unless your home happens to be a Cape Cod. Cape Cod homes have knee-walls that separate the upstairs living space from the attic area. These knee-walls allow ducts to be run in the attic and turned out at floor level in the walls. Owners of homes with knee-walls are very lucky when it comes to a multilevel heat-pump conversion.

Heat registers are typically installed in floors beneath windows set into outside walls. To get the registers in such locations, there are only a few viable ways to accomplish the goal.

If duct work is installed in an attic, supply ducts can be dropped down outside walls. This requires cutting into the finished walls to expose cavities between the framing studs. Once the ducts are installed, the channels in the walls can be repaired. While this is not major damage to existing conditions, it is much more disruptive than ceiling registers would be.

A second alternative is to cut out sections of the ceiling below the upstairs living space. By doing this, ducts can be installed between the floor joists of the second floor, much as they are under new-construction installation methods. This eliminates the need for raising the ducts through the second floor, but it leaves you with ceiling repairs to make, and it can require several duct chases on the first floor.

If you will be installing new flooring in the second-floor living space during your remodeling, you can remove sections of the second-floor subflooring, rather than removing the ceiling from below. After the ducts are installed, the flooring can be nailed back down, and the new floor covering can be installed.

With any method other than the attic method, there might be plumbing pipes, electrical wires, old radiator piping, and structural members in the path of your ducts. These complications often make it necessary to adjust your plans as you encounter them. Normally, there will be some open bays to run the duct through, but the job is rarely simple during a conversion project.

If you don't mind building small duct chases in the second-floor living space, you can drop the duct feeds from the attic to the floor in them. It will be obvious what the enclosures are for, but it is an easy way to get new duct work installed.

There is one more option that will often work with older homes. Many old houses have tall ceilings. If you have enough ceiling height, you can run ducts for the second floor and attach them to the existing ceiling joists. Once the ducts are installed under the ceiling of the first floor, you can hide them with a false ceiling. In homes with nine- or ten-foot ceilings, this method is very effective.

Electric baseboard heat

Converting from electric baseboard heat to a heat pump is not much different than the examples just discussed, except for the removal of the existing heat. The duct-work aspects will remain the same, but getting electric heat removed is much less labor intensive than the removal of radiators is.

To remove electric baseboard heat, you must make sure that the electrical power to the heating units is turned off. Even after the circuit breakers are turned off, you should check the wires with an electrical meter to make sure they are not charged with electricity. You should have the wires feeding the heaters disconnected from the main electrical box. Otherwise, someone might turn the power back on to the discontinued wires, which can create a hazard. Working inside of an electrical panel is dangerous, and I don't believe anyone except a licensed electrician should do the job.

The baseboard heating units should be screwed onto your walls. Remove the screws and the units will come right off. Cut the feed wires where they come out of the wall, install wire nuts and electrical tape over the bare wires and stuff them back into the wall. Other than for patching the small holes in the walls, that is all there is to removing electric baseboard heat.

Hot-water baseboard heat

Replacing hot-water baseboard heat with a heat pump is also similar to replacing radiators with a heat pump. However, the baseboard heating units are very light and easy to handle. There are a few types of hot-water baseboard heat that use cast-iron heating elements, and they are quite a bit heavier than the normal finned elements, but still nothing to compare with the weight of a radiator. Also the piping is usually copper, and it is considerably easier to cut than the steel pipes normally used with radiators.

As with radiators, the boiler for a hot-water heating system must be turned off and drained before the heating units can be removed. Once the boiler has been drained, the end caps on the heating units are lifted off, and you should see copper piping.

The covers for the heating elements will be screwed to the wall. Remove the screws to free the units. With that done, you will have to cut the copper piping. This is done best with what is often called a midget tubing cutter. It is a small tool that can be used in tight spaces to cut copper cleanly. A hacksaw with a metal-cutting blade will also get the job done in a hurry.

Once you have removed the heating elements and disconnected the boiler, you are ready to install duct work. The methods used are the same as described for the conversion from radiators.

Forced hot-air heat

Forced hot-air heat is usually easier to convert to a heat pump than any of the other types of heat we have discussed. The reason is simple, forced hot-air heat uses duct work that the other types of heat don't. While some of the duct work might have to be enlarged, at least there is some duct work that can be salvaged and used. It might even be possible to use most of the duct work that is already installed with the new heat pump.

The removal of a hot-air system doesn't require as much work as with the other types of systems we have discussed. All that is required is the removal of the furnace and possibly some of the duct work. The new heat pump can then be set into place and connected to the existing ducts, assuming that they are the proper size.

Unfortunately, many contractors used to skimp on the size of duct work and the number of supply registers installed. If this is the case with your home, new duct work will have to be installed, at least in some areas, and additional registers will need to be added. The worst case is having to go through the same procedures described earlier for installing duct work. The best case is that you can use most of the existing duct work without major modifications.

Insufficient duct work will cause cold spots in the home. An oversized duct system can create cold, clammy conditions when the heat pump is in its air-conditioning cycle. It is important to make sure the duct work is sized properly for the system you are installing.

Electrical upgrades

Electrical upgrades are often needed when an old house is fitted with a new heat pump. Many old electric services do not provide adequate power for the demands of a heat pump. Heat pumps require a 200-amp electrical service. Because many old homes have either 60-amp or 100-amp services, a panel-box upgrade is usually necessary. This is dangerous work and should be left to the hands of a skilled, licensed electrician.

Pad placement

Pad placement for the outside unit of a two-piece heat-pump system is not normally a problem, but it is one more construction consideration to give some thought to. Because the outside unit must be connected to the inside unit with cables and pipes, you must make arrangements to get these connections into your home. This typically means putting a hole in your foundation and providing a sleeved opening for the pipes and cables to pass through. With a brick or block foundation, this job can be done with a rotary hammer or even a regular hammer and cold chisel. If your foundation is made from thick, poured concrete, you might have to rent a small jackhammer to make the opening. Most tool-rental centers rent electric jackhammers at very reasonable rates that will get the job done.

Another consideration for the pad placement hinges on where you want the outside unit to sit. Some people dislike the appearance of the units, so you might want to spend some time finding an inconspicuous place to set it. If you live in an area with deep snowfalls, the pad should be elevated off the ground to ensure that it will not be engulfed by snow.

The biggest problem

The biggest problem with the successful completion of a heat-pump conversion is the installation of duct work. The suggestions given earlier in this chapter are all viable ways to overcome the troubles associated with installing duct work in existing, multistory homes. Each home is unique in its duct work needs and opportunities, so it is important to investigate the construction considerations for your home before making a firm commitment to convert to a heat pump.

If you take time before the conversion to plan your work, the job will go more smoothly and result in a more efficient system. Don't hesitate to confer with local professionals when it comes time to plan your installation. There is no substitute for the years of hands-on experience these professionals have. By planning your work and working your plan, you will enjoy a much easier conversion project than those who just dive into the job without knowing what obstacles to anticipate.

7

Understanding and calculating heat gain and heat loss

Understanding and calculating heat gain and heat loss is a process few consumers care to learn. To do the job properly, a person must be trained and experienced in the process. Many HVAC contractors are not able to work out heat-gain and heat-loss figures with extreme accuracy, so it would be ludicrous to assume that an average homeowner could accomplish the task easily. (The formulas and math required to compute heat gain or heat loss is presented at the end of this chapter.)

Whether you will be installing a heat pump yourself or having a contractor do the job for you, it is important to have some knowledge of heat gain. The heat gain of a house must be computed before a heat pump can be properly sized. This chapter is not going to qualify you to become a heat-gain expert, but it will give you enough knowledge to hold your own with most contractors.

The easiest way

Chapter 4 detailed the easiest way to determine your heat gain and heat loss, if you have a cooperative utility company. Many utility companies will be happy to provide heat-gain and heat-loss figures for individuals who will be building new homes. A set of blueprints and specifications on the type of building materials to be used is all that is needed for them to come up with accurate figures. It might also be possible to have a representative of the utility company come to your existing home and compute a heat gain when you are planning a heat-pump conversion.

All you have to do to find out if the utility company will help you is call. If your new heat pump will run off of electricity, call the electrical company in your area. Gas heat pumps will require you to call the local gas company. Tell the representative what your plans are, and ask if the utility provides energy-management services, such as computing heat gain.

The heat-gain figures derived from utility companies are numbers you can depend on. They are not biased to steer you into buying a particular piece of equipment or a larger unit than you need. The numbers will be accurate, and they are usually free of charge.

One rule-of-thumb method

There is one method for calculating the size of a heat pump that is not only easy, but reasonably accurate as well. This method doesn't require fancy formulas or a math degree. While the figures you arrive at will not be as accurate as those of an engineer doing a heat-gain estimate for your home, they will be close enough to put you in the right general size range.

To use this method, you must know the total square footage of living space in your home, assuming that you want to climate control all of the living space. If you have a one-story home with outside dimensions of 24' × 44', your home has a total square footage of 1,056 square feet.

To find the square footage of your home, multiply the depth by the width. In our example, the depth is 24' and the width is 44'. Twenty-four multiplied by 44 gives a total of 1,056.

In this simple heat-gain calculation, one ton of refrigeration is considered acceptable for an average of every 600 square feet of living space. Because a one-ton unit would not be large enough for the sample home, the logical choice would be a two-ton unit. The two-ton unit should handle a house with as little as 1,000 square feet or as much as 1,200 square feet.

Let's say that your home has 1,800 square feet of living space in it. What size heat pump would you need? Did you come up with a three-ton unit? If you did, you're right.

There are several factors that could influence the results of such a simple heat-gain formula, as discussed later. For average homes, under average conditions, the formula you have just learned is accurate enough to work with.

Another method

A second method of calculating heat gain is available to you from the Air-Conditioning and Refrigeration Institute (ARI). The materials from ARI are simple, easy to use, and more accurate than most rule-of-thumb methods. If you would like more information on the packages available from ARI, contact can be made at the following address:

Air-Conditioning and Refrigeration Institute
1346 Connecticut Avenue, N.W.
Washington, DC 20036

Contractors

Most contractors that sell and install heat pumps will be more than happy to provide you with a free heat-gain analysis. But, be aware that they will probably use the opportunity to get into your home and make a sales pitch. Still, if you are strong willed enough not to be sold until you are ready to be sold, contractors are a good source of free heat-gain estimates.

If you elect to have your heat-gain figured by a contractor, don't rely on just one or two contractors. Call several and compare their figures. All of the figures should come in with similar results. If there is a large spread between the numbers (either high or low) a mistake was either made or someone is trying to sell you equipment that is not the right size. This type of situation will require further investigation to determine which contractors are correct in their estimates.

Terminology

The terminology used in calculating heat gain can sound like a foreign language to people outside the trades. Here are a few of the words, terms, and phrases you might encounter when seeking the best heat pump for your home:

Absolute humidity Absolute humidity is determined by the actual amount of water vapor contained in one cubic foot of air. The measurement is expressed in grains or pounds per cubic foot. One pound is equal to 7,000 grains. Absolute humidity is equivalent to the density of air.

Humidity   Humidity is a word almost everyone has heard and used. It refers to the amount of moisture in the air. Humidity in the summer creates a muggy, uncomfortable climate that forces many people to turn on their air conditioners. A lack of humidity in winter can create static electricity in a home, dry air that is less enjoyable to breath, and other unwanted conditions. This tends to make people humidify the air in their home, either with their heating-and-cooling equipment or with a separate humidifier. Humidity, however, is too broad a term to be used when sizing a heat pump. That is why the terms *absolute humidity, relative humidity,* and *specific humidity* are used.

Relative humidity   Relative humidity is a ratio used to describe the amount of water vapor that is mixed with dry air. The process for determining relative humidity is complex and is not one that a homeowner could accomplish without special equipment and knowledge.

I'm sure you've heard weather reports give accounts of relative humidity. All you need to know for the purposes of buying a heat pump is that relative humidity is a ratio based on the percentage of dry air mixed with water vapor.

Specific humidity   Specific humidity is determined by the weight of water vapor contained in a pound of dry air. This information might seem useless to you, but it can come in handy when you are conferring with professionals.

Dry-bulb temperature   A dry-bulb temperature is a temperature we use every day. It is the temperature of air as it is measured by a standard thermometer.

Wet-bulb temperature   A wet-bulb temperature indicates at what point air will become saturated if water is added to it. For the purposes of heat pumps, a wet-bulb temperature pertains to humidity. For example, a high wet-bulb temperature suggests that the humidity is high.

Sensible heat   Sensible heat is merely the amount of heat in air that can be measured by a standard thermometer. Sensible heat, however, does not reflect the total heat needed for calculating heat gain.

Latent heat   Latent heat must also be factored in when doing a heat-gain calculation. What is latent heat? It is heat generated by the presence of people, cooking, and other daily activities. Technically, latent heat is the amount of heat contained in the moisture content of air.

Worksheets

Worksheets for calculating heat gain can be more difficult to complete than a tax return. To be done properly, the worksheets must contain information about your home that you probably don't know. For example, do you know the percentage of shading that affects your windows? Can you tell which direction is Southwest? At what latitude is your home orientated? If you know the answers to these questions, you would be a grand-prize winner in a contest against most homeowners.

A worksheet for a heat-gain calculation will contain at least three major categories: sensible heat gain, latent heat gain, and ventilation heat gain.

Sensible heat gain

Sensible heat gain will cover heat leakage, solar radiation, and internal heat sources. Heat leakage is the amount of heat that creeps into a home around windows, doors, and other openings in the home's construction. New houses should have less heat leakage than older, less-energy-efficient homes.

Solar radiation is heat that is gained from the sun. A good example would be placing your hand on a window sill and feeling how warm it is compared to a piece of wood trim that is not exposed to direct sunlight.

Internal heat sources abound in homes. The people that live in the home create internal heat. Televisions, refrigerators, cook stoves, and electric lights contribute to internal heat. All of these factors must be figured into a heat-gain equation, but there's more.

Latent heat gain

Latent heat gain is generated by people, ventilation, and the infiltration of heat. Again, all of these categories must be rated on the worksheet.

Ventilation heat gain

Ventilation heat gain is made up of sensible heat and latent heat that is present due to outside sources. All homes require a certain amount of ventilation to remove objectionable odors and to maintain comfortable living conditions. For example, your home might be rated for ten air changes per hour. This means that the inside air is removed and replaced ten times each hour. The recommended number of air changes could be less, or it could be much higher. You can imagine that an average person has little chance of factoring in all the possibilities for heat gain.

Actual calculations

Actual calculations for computing heat leakage, solar gain, internal gain, and ventilation gain are complex, to say the least. Let me give you an example of how to figure solar gain.

If you want to know what the solar gain for your home is, and you must if you want the heat pump to be sized properly, you should use the following formula:

The solar gain is equal to the area of all window glass multiplied by a factor for the amount of shading provided for the glass, multiplied by the heat gain by solar radiation, multiplied by the heat gain by convection and radiation.

Do you think you can arrive at an accurate solar-radiation figure? Well, you would need prescribed figures, like those found on heat-gain worksheets to factor into the equation, but you could probably do it. The point is, why would you want to go to so much trouble when there are so many easy ways to have professionals do it for you, at no cost to you?

If you really want to do your own heat-gain calculations, you should contact the American Society of Heating, Refrigeration, and Air-conditioning Engineers (ASHRAE). They have all the charts and tables needed to make accurate computations on heat gain. You can contact ASHRAE at the following address:

ASHRAE
1791 Tullie Circle, N.E.
Atlanta, GA 30329

I've been talking in terms of heat gain, because that is the factor used to determine the size of a heat pump. When thinking of heating the home in winter, people tend to think of heat loss, rather than heat gain. And that's all right. The same basic principles apply to both heat gain and heat loss. The big difference is the direction in which the loss is traveling. Heat gain has heat coming into a home that you are trying to cool. Heat loss occurs when heat escapes a home that is being heated. See Table 7-1 for a comparison of the heat-storing abilities of several materials.

Table 7-1 Heat storage comparisons

Material	Heat storage (Highest numbers are best)
Water	9
Wood	8
Oil	7
Air	6
Aluminum	5
Concrete	4
Glass	4
Steel	3
Lead	2

Improving your heat gain-ratio

Improving your heat-gain ratio is something that you can do to save money. How can you reduce the amount of heat gain in your home? There are many ways of doing it. The major sources of heat gain in buildings include:

- windows
- doors
- outside walls
- partitions between heated and unheated space
- ceilings
- roofs
- uninsulated wood floors between heated and unheated space
- air infiltration through cracks in construction
- people in the building
- lights in the building
- appliances and equipment in the building.

Many of the common sources of heat loss in buildings include:

- windows
- doors
- outside walls
- partitions between heated and unheated space
- ceilings
- roofs
- concrete floors
- uninsulated wood floors between heated and unheated space
- air infiltration through cracks in construction.

Table 7-2 shows the heating load form of a specific heat pump. Table 7-3 shows a cooling load estimate form.

Table 7-2 Heat load form. Friedrich Air Conditioning Co.

		°F. TEMP. DIFFERENCE
WALLS: (Linear Feet)		
2" Insulation	Lin. Ft. x 1.6	
Average	Lin. Ft. x 2.6	
WINDOWS & DOORS: (Area, Sq. Ft.)		
Single Glass:	Sq. Ft. x 1.13	
Double Glass:	Sq. Ft. x 0.61	
INFILTRATION - WINDOWS & DOORS: AVG.	Lin. Ft. x 1.0	
Loose	Lin. Ft. x 2.0	
CEILING: (Area, Sq. Ft.)		
Insulated (6")	Sq. Ft. x 0.07	
Insulated (2")	Sq. Ft. x 0.10	
Built-up Roof (2" insulated)	Sq. Ft. x 0.10	
Built-up Roof (½" insulated)	Sq. Ft. x 0.20	
No Insulation	Sq. Ft. x 0.33	
FLOOR: (Area, Sq. Ft.)		
Above Vented CrawlSpace		
Insulated (1")	Sq. Ft. x 0.20	
Uninsulated	Sq. Ft. x 0.50	
* Slab on Ground	Lin. Ft. x 1.70	
1" Perimeter insulation	Lin. Ft. x 1.00	
* Based on Linear Feet of outside wall	TOTAL HEAT LOSS PER °F	BTU/HR/°F

Multiply total BTU/HR/°F x 30 and plot on graph below at 40°F. Draw straight line from 70 base point thru point ploted at 40°F. Intersection of this heat loss line with unit capacity line represents the winter design temperature in which the unit will heat the calculated space.

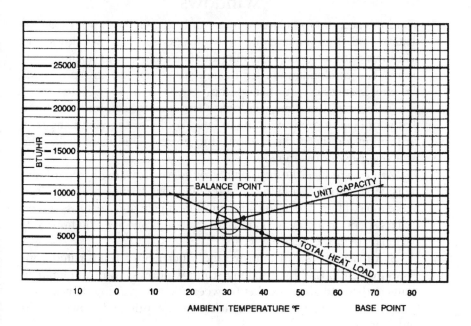

Table 7-2 Continued

HEAT LOAD FORM

The heat load form may be used by servicing personnel to determine the heat loss of a conditioned space and the ambient winter design temperatures in which the unit will heat the calculated space.

The upper half of the form is for computing the heat loss of the space to be conditioned. It is necessary only to insert the proper measurements on the lines provided and multiply by the given factors, then add this result for the total heat loss in BTU/Hr./°F.

The BTU/Hr. per°F temperature difference is the 70°F inside winter designed temperature minus the lowest outdoor ambient winter temperature of the area where the unit is installed. This temperature difference is used as the multiplier when calculating the heat loss.

The graph shows the following:

Left Hand Scale	Unit capacity BTU/Hr. or heat loss BTU/Hr.
Bottom Scale	Outdoor ambient temperature, base point.
Heat Pump Model	BTU/Hr. capacity heat pump will deliver at outdoor temperatures.
Balance Point	Maximum BTU/Hr. heat pump will deliver at indicated ambient temperature.

Below is an example using the heat load form:

A space to be conditioned is part of a house geographically located in an area where the lowest outdoor ambient winter temperature is 40°F. The calculated heat loss is 184 BTU/Hr./°F.

Subtract 40°F (lowest outdoor ambient temperature for the geographical location) from 70°F (inside design temperature of the unit) for a difference of 30°F. Multiply 184 by 30 for a 5500 BTU/Hr. total heat loss for the calculated space.

On the graph, plot the base point (70°) and a point on the 40°F line where it intersects with the 5500 BTU/Hr. line on the left scale. Draw a straight line from the base point 70 thru the point plotted at 40°F. This is the total heat load line.

Knowing that we have a 5500 BTU/Hr. heat loss, and we expect that our heat pump will maintain a 70°F inside temperature at 40°F outdoor ambient, we plot the selected unit capacity BTU/Hr. of the unit between 35° and 60° on the graph and draw a straight line between these points. Where the total heat loss line and the unit capacity line intersect, read down to the outdoor ambient temperature scale and find that this unit will deliver the required BTU/Hr. capacity to approximately 30°F.

Windows

Windows account for a large amount of heat gain, both from solar radiation and from infiltration. In winter, loose-fitting windows and windows without good insulating qualities allow cold air into your home. The combination of letting cold air in and hot air out reduces the effectiveness of your heat pump.

In summer, the same windows allow heat to leak into the home that your heat pump is working to keep cool. Consequently, the cool air from inside the home is allowed to escape. This again makes the heat pump work overtime.

You can reduce the heat gain around windows with some inexpensive methods. Caulking around window frames and sills can make a big difference in the amount of heat gain and heat loss. A few dollars spent on a couple of tubes of caulking can save you hundreds of dollars over the next few years.

Storm windows can improve the efficiency of uninsulated windows tremendously. By installing storm windows, you can eliminate drafts and create an air pocket between the two window panes. This type of work will pay for itself quickly if the existing windows are old and uninsulated.

Table 7-3 Cooling load estimate form. Friedrich Air Conditioning Co.

HEAT GAIN FROM	QUANTITY	FACTORS DAY				BTU/Hr. (Quantity x Factor)
		No Shades*	Inside Shades*	Outside Awnings*	(Area x Factor)	
1. WINDOWS: Heat gain from sun.						
Northeast	___ sq ft	60	25	20	Use	___
East	___ sq ft	80	40	25	only	___
Southeast	___ sq ft	75	30	20	the	___
South	___ sq ft	75	35	20	largest	___
Southwest	___ sq ft	110	45	30	load.	___
West	___ sq ft	150	65	45	Use	___
Northwest	___ sq ft	120	50	35	only	___
North	___ sq ft	0	0	0	one.	___
	* These factors are for single glass only. For glass block, multiply the above factors by 0.5; for double glass or storm windows, multiply the above factors by 0.8.					
2. Windows: Heat gain by conduction. (Total of all windows.)						
Single glass	___ sq ft		14			
Double glass or glass block	___ sq ft		7			___
3. WALLS: (Based on linear feet of wall.)		Light Construction		Heavy Construction		
a. Outside walls						
North exposure	___ ft	30		20		
Other than North exposure	___ ft	60		30		___
b. Inside Walls (between conditioned and uncondi- tioned spaces only)	___ ft		30			___
4. ROOF OR CEILING: (Use one only.)						
a. Roof, uninsulated	___ sq ft		19			
b. Roof, 1 inch or more insulation	___ sq ft		8			___
c. Ceiling, occupied space above.	___ sq ft		3			___
d. Ceiling, insulated with attic space above	___ sq ft		5			___
e. Ceiling, uninsulated, with attic space above	___ sq ft		12			___
5. FLOOR: (Disregard if floor is directly on ground or over basement	___ sq ft		3			___
6. NUMBER OF PEOPLE:	___		600			___
7. LIGHTS AND ELECTRICAL EQUIPMENT IN USE	___ watts		3			___
8. DOORS AND ARCHES CONTINUOUSLY OPENED TO UNCONDITIONED SPACE: (Linear feet of width.)	___ ft		300			___
9. SUB-TOTAL	x x x x x	x x x x x				___
10. TOTAL COOLING LOAD: (BTU per hour to be used for selec- tion of room air conditioner(s).)	___ (Item 9) x ___ (Factor from Map) = ___					

The best defense against heat gain and heat loss around windows in old homes is the replacement of the old windows. New, energy-efficient windows with double, or triple, panes of insulated glass are ideal. Unfortunately the work and expense involved with installing replacement windows is beyond the reach of many homeowners who are working on tight budgets.

Doors

A lot of air can leak in around doors that are not sealed properly. Adding weatherstripping and sill seals to doors can reduce drafts and save you money on your heating and cooling costs.

Replacing a door or two is much less complicated, and less expensive, than replacing a house full of windows. With modern insulated doors, you can again improve the efficiency of your heat pump.

Insulation

Having proper insulation in your home is very important if you want to keep your heat gain figures at a minimum. Most old homes are very deficient in their insulation requirements. Some homes have no insulation. Others have a little insulation in the walls and none in the attic. Until you inspect the insulation of an older home, there is no way to know what is there, or what condition the insulation is in.

There should be insulation under the first floor of a home, unless the basement is heated. Insulation should be in all exterior walls, and in the attic. The amount of insulation needed in these areas varies from region to region. A phone call to your local building inspector will get you the minimum insulation requirements in your area.

Adding insulation under the first floor of a home or in an attic is easy. The work can be done usually by anyone who is in good health and who is not allergic to the insulation. Adding insulation in outside walls is not so easy, but it can be done.

Wall insulation is usually added to existing homes by removing siding or drilling holes in the siding and blowing insulation into the wall cavities. This job should normally be left to professionals, but it is possible for handy homeowners to accomplish the task. (See Tables 7-4 through 7-8.)

Table 7-4 Approximate mineral wool insulation needed when blown in with a machine

R-11	3.5
R-19	6
R-22	7
R-30	10
R-38	12.5
R-49	16

Table 7-5 Approximate glass fiber insulation needed when blown in with a machine

R-11	5
R-19	8.5
R-22	10
R-30	13.5
R-38	17
R-49	22

Table 7-6 Approximate glass fiber insulation needed in the form of rigid boards

R-11	3
R-19	5
R-22	5.5
R-30	7.5
R-38	9.5
R-49	12.5

Table 7-7 Approximate cellulose insulation needed when blown in with a machine

R-11	3
R-19	5.5
R-22	6
R-30	8.5
R-38	10.5
R-49	13.5

Table 7-8 Approximate glass fiber batt insulation needed

R-11	3.5
R-19	6
R-22	7
R-30	10
R-38	12.5
R-49	16

The benefit of upgrading insulation, windows, and doors will be enjoyed for many years to come as you pay the fuel bills for your new heat pump. Doing these upgrades might even allow you to buy a smaller heat pump. When this is the case, you save two ways. Money is saved on the purchase of the heat pump, and the cost to operate the heat pump will be less for the years to come.

8

Sizing your equipment

Sizing your heat pump equipment is a job that you should probably leave up to full-time professionals. The size of a heat pump is critical to its satisfactory operation, and few homeowners have the knowledge needed to size a system properly. Even if you plan to do all of the installation work for your heat pump personally, you should strongly consider asking for professional help in sizing and designing the system.

An HVAC professional can look over load calculations for heat loss and heat gain and determine the size of a unit you need without much trouble. A homeowner, on the other hand, has two problems when trying to size a heat pump. The first obstacle is doing an accurate load calculation for the home. A second hurdle is met when the homeowner must try to interpret the load calculation and size a heat pump. The last chapter showed you the basics of heating and cooling load calculations. Now it is time to take the process a step further and size a heat pump.

When you size a heat pump, you must make sure you don't select a size that is too large or too small. The size of a heat pump is rated in tons. Most residential homes use heat pumps that range in size between two tons and four tons. There are exceptions, of course, but as an average, you can expect your work to point in the direction of a two- to four-ton heat pump.

Heat-only heat pumps

Heat-only heat pumps are sometimes used in residential applications, however, most heat pumps provide heating and cooling for homes. It is most likely that you are considering a heat pump that will not only heat your home, but provide air conditioning as well. There are, however, some people who only want their heat pumps to produce heat. Before you can size your unit, you must know if it will be a heat-only heat pump or a heat pump that provides both heating and cooling.

If you are interested in a heat-only heat pump, size it based on heat loss and a winter heating load. You have two options when you get ready to size this type of unit. Some professionals believe the heat pump should be sized to meet the demands of the coldest winter load. Other professionals think it is best to size the heat pump to meet up to 85% of the winter load and to allow the back-up heat to make up for the missing 15% of the highest load.

Why would a professional recommend what appears to be an undersized heat pump? The heat pump will only be undersized on the most severe winter days. These days do not account for much of the heating season. Therefore, a smaller heat pump will run more efficiently and at lower operating costs at all times, except those short periods of time when the outside temperatures are extreme. This makes sense when you give it enough thought.

The decision between sizing the heat pump to a full capacity or to the slightly undersized capacity must be weighed and made by you. It is likely that the severe temperatures of the winter design should account for less than 3% of the heating season. By biting the bullet on increased costs for back-up heat during that 3% of the season, you can save money for the remainder of the heating period.

The important thing to remember if you are sizing a heat-only heat pump is that you will be calculating its size based on a winter design load. This is a point that could be easily overlooked because most heat pumps are sized with a summer design load.

Combination heat pumps

Combination heat pumps cool buildings during the summer and warm them in the winter. This type of heat pump is, by far, the most common type in residential use. Unlike heat-only heat pumps, combination heat pumps are sized to meet their air-conditioning needs. To do this, the person sizing the system must pay attention to heat gain and summer design temperatures.

If a heat-pump system is not sized properly, one of two problems will arise. A heat pump that is too small simply cannot perform its job satisfactorily. Heat pumps that are too large don't perform well in their air-conditioning mode, and they waste energy. Finding suitable middle ground between heating performance and air-conditioning requirements can be tricky business.

It would seem logical that in cold regions, like the state of Maine, heat pumps should be sized to heat homes effectively. The fact that air conditioning is only a small part of a heat pump's job in such climates would tend to support this theory. In truth, however, a heat pump that provides both heating and cooling should always be sized for its cooling performance. This calls for some compromise in the heating requirements, but it is the proper way to size a combination heat pump.

Let's talk a little about the compromise between heating and cooling. To set the stage for our discussion, let's say we will discuss the sizing considerations for a home in Maine, a home in Virginia, and a home in Georgia. Each of these homes is going to be fitted with a combination heat pump. This means the size of the heat pumps will be determined by the demands for cooling, rather than heating. Let's start with the house in Georgia.

Georgia is, for the most part, a warm region. The state gets some snow and some cold temperatures, but if you lived in Georgia, air conditioning would probably be of much more concern to you than heating, at least in terms of how often it would be used for extremes. Assuming that you were the homeowner in Georgia, you would not have any qualms about sizing your heat pump based on summer design temperatures. With your system sized for the summer requirements, heating your home to a comfortable temperature in winter would not be a problem. But, what about the house in Virginia?

If you were sizing a new heat pump for your home in Virginia, you might be a little more concerned about its heating ability. There is little doubt you would still have a strong interest in the heat pump's cooling ability, but the heating aspect of the equipment might weigh more heavily on your mind. The winters in Virginia are not brutal, but they can produce sustained cold temperatures that are below the balance point of many heat pumps. Does this mean your home is going to be cold? No, because the back-up heat is going to pick up the slack on the colder occasions.

In Virginia, the balance of importance between heating and cooling might be nearly equal. There are plenty of hot days during the summer, and more than enough cold nights during the winter. The climate, however, is moderate enough to allow most people to feel comfortable with a heat pump that has been sized by using summer design temperatures. Now, what about the house in Maine?

If you were sizing a heat pump for a house in Maine, the major concern would be the heating mode of the unit. While air conditioning might be used a few weeks out of the year in Maine, it is not nearly as important as the heating capability of the heat pump. Even so, the heat pump must be sized based on its cooling requirements. I know this seems strange, but it's true. If you were to size the heat pump using winter design temperatures, the system would be far to large for proper cooling of the home.

There is a way to cheat a little on the sizing of a heat pump for a cold climate. Typically, it is well accepted that a heat pump can be oversized by as much as one-fourth its rated size without losing performance potential. In a situation such as the one for the house in Maine, it would be logical to boost the size of the unit to offset the cold winter temperatures. Even upgrading the size of the unit will not eliminate the need for back-up heat in extremely cold temperatures, but it will reduce the frequency with which the auxiliary heat is needed.

The balance point

When you talk about sizing a heat pump, you must talk about the balance point. This is a theoretical point on a graph where a heat pump loses its ability to perform well. Outside temperature is the trigger for the balance point. When a heat pump is in the heating mode, it will have to work harder as the outside temperature decreases.

To establish a balance point, there are a few factors that must be used in conjunction with each other. First, you must determine what outside temperature will represent a point where heat is not called for. Many professionals use a temperature of 65 degrees to represent the no-load temperature. The second factor that must be established is the BTU requirement, per hour, to heat a home. When you know these two numbers, you can fashion a chart that will help you to size your heat pump.

Assume that 65 degrees is your no-load temperature. Further assume that your design heating load calls for 40,000 BTUs per hour at zero degrees F. On a piece of graph paper, draw a vertical line on the left side of the paper. Along the line, at different intervals on the paper, mark numbers to indicate the BTUs needed per hour. Start with zero and work your way up to 40,000. Make your marks in increments of 10,000 BTUs. In other words, your first number is zero, the second number is 10,000, the third number is 20,000, the fourth number is 30,000, and the final number on the vertical line is 40,000.

Now draw a horizontal line that intersects with the base of the vertical line. The horizontal line should run from left to right. This horizontal line is going to represent the outside temperature range. Start where the two lines connect, and write in a temperature of minus 10 degrees. Move to your right and mark in a zero temperature.

Continue moving to the right, at regular intervals, and adding temperature readings in 10-degree increments. Stop when you get to 70 degrees.

With your chart complete, take a pencil and make a mark at the 65-degree location on the horizontal line, this will be your no-load temperature. With the help of a ruler, draw a diagonal line from the no-load temperature towards the 40,0000 BTU mark on the vertical line. If your chart was to scale, you would see that your balance point would be approximately 27 degrees. The chart you've just drawn might not be that accurate, but you now know what a balance-point chart looks like. The actual balance point will depend on one piece of information that we didn't use, the performance of your particular heat pump. I think, however, that you will find most average heat pumps will have a balance point of around 27 degrees. Anything less than 30 degrees is normally accepted as being good.

Choosing your balance point might be affected by the cooling load of the unit. If you recall, a heat pump should not be oversized for cooling by more than 25%. When you are thinking in terms of heating, you want the lowest balance point you can get, but you might not be able to have it.

When a heat pump is being sized for both heating and cooling, the predominant factor in the size of the unit will be the cooling load. This can influence the balance point you are seeking. For example, assume you were working with three different heat pumps. The first heat pump you sized will work fine for meeting the cooling load, however, its balance point is up near 40 degrees and is not suitable for heating effectively in the winter.

When you rate out a second heat pump, you find that it will easily meet your cooling needs, and its balance point is around the freezing point. This unit would work for both your heating and cooling needs, but you'd like a balance point that is below 30 degrees. With this in mind, you do a worksheet on a third heat pump. Its balance point is 27 degrees, and that's great for your heating needs, but there is a problem. This unit exceeds the limit of 25% oversizing for your cooling requirements.

You now have one heat pump that would be great for heating and not so good for cooling. There is also a heat pump that would do fine for cooling your home, but its balance point is too high for serious heating. Your only other choice is the heat pump with a balance point of 32 degrees. Given these three choices, you must compromise on your heating needs and go with the heat pump that has the 32-degree balance point.

Any balance point that is determined from a chart is unlikely to be the actual operating balance point of the heat pump. Many factors come into play when judging the true balance point of a heat pump after it is installed and working. For example, if you came up with a balance point for your cooling mode of 32 degrees, the actual balance point could be 30 degrees or 34 degrees. The charts used to establish balance points don't take all aspects of an operating heat pump into consideration.

A house receives a certain amount of heat through solar gain. When the sun shines through a window, heat is accumulated in the home. This type of heat is not accounted for on a balance-point chart, and therefore, the true balance point of a heat pump in the field is likely to be somewhat different than projections made on paper. Other factors that influence the true balance point include installation procedures for the heat pump and its system of ducts, maintenance (such as clean filters), local weather patterns, and more. As long as you find your heat pump to have a balance point of less than 30 degrees, it should do fine. Keep in mind, of course, the possible need to compromise in order to meet cooling needs.

How much is a ton?

A ton is usually thought of as 2,000 pounds. However, when talking in terms of heat pumps, a ton is equal to 12,000 BTUs per hour. If you have a three-ton heat pump, it is rated at 36,000 BTUs per hour. If you go back to the example used in discussing balance points, the heat pump we talked about had a heating design load of 40,000 BTUs. What size heat pump would be needed to cover 40,000 BTUs per hour? The proper size for the heat pump would be three and a half tons.

Once you know the design temperature of a heat pump, figuring its size is easy. Take the design temperature and divide it by 12,000. The result will be the size of the heat pump you need. If your design temperature is 24,000 BTUs per hour, you need a two-ton heat pump.

Supplemental heat

Most heat pumps are equipped with some amount of supplemental heat. The exception to this would be heat pumps in very mild climates, such as Florida, where outside temperatures don't drop below the unit's balance point. You already know that a good balance point is often somewhere between 25 and 30 degrees. For the sake of the example we are about to explore, assume that your heat pump has a balance point of 27 degrees.

When the outside temperature drops below the balance point of your heat pump, you simply will not get enough heat production out of the heat pump to maintain an average indoor temperature of 70 degrees. This is when you need supplemental heat. The auxiliary heat is normally of an electric resistance type, and it can be expensive to operate. That is why the lower the balance point is, the better off you are, unless it causes problems with your cooling load.

How much supplemental heat should your heat pump have? It depends on your winter design temperatures, local code require-ments, and your comfort level. If you don't mind your home having a cool indoor temperature, of say 65 degrees, you won't need as much auxiliary heat as a next-door neighbor who requires a room temperature of 72 degrees. However, unless you live somewhere where heat is not important, you should have some back-up heat.

The back-up heat in your heat pump might have to meet certain minimum code requirements. Because heat pumps can fail, the elec-tric heat (or some other type of supplemental heat) could be your only heat during periods of failure on the part of the heat pump. Without emergency back-up heat, your water pipes could freeze, and you could get downright uncomfortable if the heat pump was out of commission for long. Check with your local code enforcement office to see if there are any specific requirements in your area that pertain to the amount of auxiliary heat that is required when a heat pump is installed.

Going back to our example of needing 40,000 BTUs per hour to maintain a comfortable indoor temperature during winter design temperatures, we can investigate the amount of supplemental heat that might be desirable. Perhaps the easiest way to discuss this would be to talk about the desired amount of back-up heat in terms of per-centages. Let's say that you enjoy an indoor temperature of 70 de-grees. You want enough back-up heat to keep you and your pipes from freezing if something goes wrong with the heat pump. After a little thought, you decide that an inside temperature of 50 degrees would be a little on the cool side during the winter, but it would cer-tainly be warm enough to protect both you and your plumbing. What percentage of 70 degrees is 50 degrees? In round figures, its about 30%. Therefore, you would need 70% of your normal BTU perfor-mance to give you an indoor temperature of about 50 degrees. If you take your normal BTU rating of 40,000 BTUs per hour and deduct 30% of it, you will arrive at a number of 28,000 BTUs per hour. This means that you would want back-up heat rated to give you 28,000 BTUs per hour.

Of course, you can adjust the level of back-up heat installed with your heat pump to suit your personal desires, so long as there are not code requirements that make you do otherwise. Some people might gamble and settle for 40% of their normal temperature while others wouldn't feel safe with less than 80% of their heating requirement covered by the back-up heat. Just work the numbers, on the percentage basis, and you can easily determine how much auxiliary heat is enough for you.

Consult professionals

The most important piece of advice I can give you in terms of sizing your heat pump is that you should consult professionals. The sizing of the unit is of utmost importance. If a mistake is made in the sizing process, you could live a long time in an uncomfortable house regretting your decision to avoid the help of professionals.

I am a professional contractor, but when it comes to figuring heat gain and heat loss and sizing a system, I depend on engineers to come up with the right numbers. There are so many factors that must be calculated with precise accuracy during a load calculation that anyone who is not accustomed to doing such work on a regular basis can have difficulty. And I'm not talking just about homeowners, even full-time professional installers can have a lot of trouble sizing a heat pump within tight standards. It is fairly easy to use a ballpark figure to determine what size heat pump is needed, but estimated figures can hurt the user of the heat pump over the long haul.

Most companies that sell heat pumps are more than happy to do your load calculations and sizing for you; let them. If I were going to install a new heat pump in my home, I'd consult at least two suppliers, and probably three. My request would be for them to compute the load calculations for my home and recommend the proper size for the heat pump. Once I had the results of their work, I could compare the recommendations and see how close they were. If all of the estimates were close in their numbers, I'd believe them. However, if there was a significant difference in the recommendations, I'd pursue the cause for the discrepancy. Did someone make a mistake? Is one supplier trying to sell me more heat pump than I need? Who's right? These are the types of questions that would run through my mind. If necessary, I'd go to more suppliers and get more estimates until I could draw a conclusion of what was going on.

While I strongly advise you to take advantage of professional help in sizing your equipment, I will give you information on a publication that will help you do the load calculations yourself. The Air Conditioning Contractors of America (ACCA) offer a publication titled *Manual J.* The manual contains 126 pages of text, tables, and charts that are used to calculate heating and cooling loads. If you would like a copy of the manual, it can be purchased from ACCA at the following address:

Air Conditioning Contractors of America
1513 16th Street, NW
Washington, DC 20036
202-483-9370 (Phone)

ACCA has a comprehensive line of publications that might be of use to you in your quest to install a new heat pump. The publications are basically intended for professional use, but there is a lot of good information in them that an average person can use.

9

Air-flow considerations

There are many air-flow considerations to be addressed when planning for a new heat pump. You must determine how the duct work can be run to make air distribution easy and effective. Finding a suitable place for the cold-air return will be one of the jobs involving airflow considerations. Deciding where to place floor registers and how to install the duct work will be another.

Air flow is critical to the satisfactory operation of a heat pump. If the air flow is not right, the heating-and-cooling system will not function properly. The problem is, planning and installing the perfect duct system is not easy, even for professionals. A lot of thought must go into the routing and installation of the duct work before any on-site work begins.

It is unlikely that the average homeowner will be able to do all the design and installation work required for a heat-pump installation without some professional help. Don't be frustrated by this. The more you can do yourself, the more money you can save. And even if you aren't going to do any of the work, you will at least be able to talk intelligently with the professionals doing the work for you.

Basic rules of good air flow

There are some basic rules to observe when working on a layout for duct work. These facts will be helpful to you whether you are contracting the work out or doing it yourself.

Low velocity

A low velocity of air distribution does not make as much noise as a high-velocity movement. Lower velocity can also translate into lower operating costs.

It is not unusual for commercial work spaces, like factories, to use higher velocities in their air distribution than what would be used in a church, school, or residence. Because factories tend to have high-noise levels from the work being done, the sound of air moving through ducts quickly is not so distracting as it would be in other settings.

The right size

Having ducts that are the right size for the job being done is very important. If the ducts are too big or too small, the heat pump cannot produce ideal air distribution. This problem typically arises when existing hot-air furnaces are replaced with heat pumps. The ducts used with hot-air furnaces are normally too small to provide the best service possible from a heat pump.

Keep it short

The best way to make duct work perform well is to keep it short. The shorter the distance air must travel is, the better the results of heating and cooling will be.

Some people install duct systems where the supply ducts must run from the center of a home to the far end of the house. This is not the right way to do the job. For proper air distribution, a main trunk line should be installed to carry air to the supply ducts. By installing the system in this manner, supply ducts can be kept short and efficient. The following is an example of a house I once rented that had its duct work installed wrong.

The house we lived in had a forced hot-air furnace. The furnace was set in the basement, near the front wall of the home. There was very little trunk line run and what was there, was not sized properly. Trunk lines should get smaller as they get longer. The trunk line in the house we rented was the same size for its entire distance.

Whoever installed the duct work had run small, round ducts off of the trunk line. Some of these supply ducts covered distances of nearly 20 feet. Not only was this bad, the installer did not place floor registers in a consistent manner.

Our bedroom was on the front, north corner of the home, and it had only one supply outlet. The round duct ran about twenty feet from the furnace. The room had three windows in it and only one floor register. It should have had a register under each of the windows, and the supply ducts should have come off of a trunk line that was much closer to the room. Our bedroom was always cold in the winter.

The bedroom our daughter used had two windows and two floor registers. Her room was also on the north side of the house, but it was much closer to the furnace and trunk line. With the thermostat in the hall set at 68 degrees F, I could not stand to stay in my daughter's room with the door closed. It was too hot.

The downstairs bathroom had one heat register, and again, it was too hot to be comfortable, even with the door open. The bathroom was positioned directly above the trunk line and near the furnace.

The living room was large and contained a lot of windows. There were two windows on the front wall, two windows on the side wall, and a picture window with two smaller windows beside it on the back wall. With all the glass and wall exposure, there were only three heat registers. The room was comfortable in my estimation, but my wife and daughter often complained of the cold drafts and chilly temperatures.

The kitchen and dining room were set off behind the main house. These rooms were the farthest away from the furnace. Both rooms contained a lot of glass on the outside walls, and there were three heat registers to serve both rooms. This area stayed so cold that it was uncomfortable to eat in the dining room, and walking on the floors in stocking feet was a chilling experience.

Then there were the two rooms upstairs. These rooms were large, and each room contained four windows. One of the rooms had two heat registers and the one I used as my office had only one. Even with the thermostat set on 70 degrees F, my wife would have to run an auxiliary heater to work in my office.

When we first moved into that house, we used the large rooms upstairs as bedrooms. The cold Maine winters and the inefficient duct work quickly changed our thinking about where the bedrooms should be located, so we moved them downstairs.

The furnace in the home was large enough to work fine, but the duct work and air distribution was so poor that the furnace didn't have a chance. It was possible to be comfortable in one room, too hot in another, and too cold in another, all at the same thermostat setting in the hall.

It is a shame that the duct work in that house was installed as the house was being built and that there was no excuse, except incompetence on the part of the installer, for creating such an inefficient heating system. However, this example illustrates clearly what can happen if proper attention is not paid to air distribution.

Changes in direction

When installing duct work, changes in direction can have direct, and adverse, affects on the air distribution. All changes in direction should be kept to an absolute minimum. When offsets are required, the change in direction should be made with as much of a sweeping motion as is possible. For example, a 45-degree offset will not disrupt the air flow as much as a 90-degree bend.

Register locations

Register locations are another important part of obtaining suitable air distribution. As a rule of thumb, there should be a floor register under every window installed in outside walls. In some few cases this might be overkill, but for the average house, putting a register under each window will work well.

For heating purposes, duct outlets should be kept low to the floor. This might mean having the register in the floor, or it could involve mounting a wall register close to the floor. Heat rises, and duct outlets in ceilings do not distribute warmth as well as floor registers.

Ceiling registers work fine when the heat pump is being used to cool a home, and they can be used to disperse heat. In fact, commercial buildings frequently have their registers mounted in ceilings. Common practice, however, when installing ducts in residences calls for floor-mounted registers.

Resistance

Resistance in duct work will block and disrupt uniform air distribution. For duct work to convey air effectively, resistance in the ducts should be kept at a minimum. This can be done with thoughtful planning of how the duct work will be installed to avoid bends and turns.

There are two types of pressure losses caused by resistance encountered in duct work. The resistance created by sharp turns is called *dynamic loss of pressure*. Dynamic pressure loss can be kept to a minimum by avoiding offsets in the ducts.

The second type of pressure loss is called *friction loss*. The resistance in this case is the wall of the duct work. A smooth interior in duct work results in less friction loss than a rough surface would.

When duct calculations are being made, these resistance factors must be allowed for. To arrive at proper duct sizing, the total static pressure drop (created by resistance) must be ascertained. The resistance inside of duct work is referred to as *external static pressure*. This might seem strange because it would be logical to assume the losses of pressure within duct work would be considered internal losses. The reason it is called an external loss is because the loss is occurring in the duct work, rather than in the heating unit.

Calculating internal pressure losses can be done with the help of tables supplied by manufacturers. To calculate external pressure losses, one must have the appropriate rating tables to work with. These tables are available from ASHRAE.

Blowers

When blowers are sized for a heating system, their size is determined by dealing with both the external static pressure losses and the internal static pressure losses. The blower selected must be powerful enough to overcome both types of pressure losses. Otherwise, the air cannot be circulated properly.

Air outlets

The air outlets of a duct system are usually called either *grilles* or *registers*. These are the openings where air from the heating or cooling system is emitted. There are also outlets called *diffusers*.

Grilles

Grilles are typically used when the air outlets are mounted in walls. The grilles are equipped with some type of directional control apparatus. By giving the user of a system ways to direct the air flow, grilles allow for more versatility in air flow than a open grate would.

Registers

Floor registers are more common than grilles. The reason for this is simple; more air outlets are installed in floors than in walls. Registers are very similar to grilles, but they do posses one major difference. The volume of air flow from a register can be controlled. Grilles allow the directional movement of air, but registers allow directional movement and the option of controlling the amount of air allowed to escape from the duct system.

If it is desired, registers can be closed, by way of a damper, to eliminate nearly all of the air flow from a particular supply duct.

There are times when this feature is very desirable. Let me give you an example of when I close one of my floor registers completely.

I have a young daughter, Afton, who gets very excited about Christmas. Usually, right after Thanksgiving, Afton starts asking when we can put up our Christmas tree. My wife, Kimberley, and I try to put her off until the second week of December. We are not always successful.

Because we always use fresh-cut trees, there is the problem of keeping the trees from drying out. This is especially true because we tend to leave the tree up until New Year's Eve.

The location we use for the tree has two heat ducts nearby, one on either side of the tree. If the registers for these supply ducts were left open, the dry heat would sap moisture from the tree, creating a fire hazard. Not wanting the tree to dry out prematurely, I always close off the two heat registers while the tree is in the house.

In this case, the reason for closing the registers is to keep a Christmas tree from drying out, but there are other reasons for wanting to close a register. If a room is not being used, such as when a child is away at college, the registers to that room can be closed to conserve heat. If a piece of furniture is being placed over a register, some people will choose to close the damper. Registers give you that little extra control over air flow that can come in handy.

While registers are normally installed in floors, they also can be used in wall locations. Not all registers allow the movement of their face plates, but some do. If you purchase registers that have adjustable face plates, to control the direction of flow, you will have the best of both worlds.

Diffusers

Diffusers are used to direct air flow. A diffuser can take on many shapes. Some are shaped to sit over floor registers and direct air flow at a 90-degree turn. Rather than having air from the duct blown up towards a ceiling, the diffuser forces the air to be blown out over the floor. Extended versions of this type of diffuser are available for getting air flow from a floor register out from under a piece of furniture.

If a sofa were to be set over a floor register, the air flow would be disturbed, and the sofa might become too hot. By using an extended diffuser, the air flow can be directed, through the diffuser, from the floor register to a point at the edge of the sofa. This allows the air from the duct to enter the room unobstructed by the sofa.

Diffusers are also used on ceiling outlets. A cone-shaped diffuser concentrates the air flow coming out of the register or grille and forces it downward. This use is effective, but the cone-affect can create hot spots (or cold spots if air conditioning is being used) that are uncomfortable to be around.

Dampers

Dampers can play an important role in gaining proper air distribution. With the use of dampers, it is possible to direct air to certain areas, reduce the amount of air flow in a duct, or stop air flow in a duct. This type of control allows users to modify air flow to meet their specific needs. When you think of dampers for heating and cooling systems, there are two basic types to consider; they are *volume dampers* and *splitter dampers.*

Volume dampers

Volume dampers are more common than splitter dampers and squeeze dampers. Air flow in a duct can be modified or cut off completely with the use of a volume damper. When talking about volume dampers, there are three types of them to give serious thought to. The three types are butterfly dampers, slide dampers, and hit-or-miss dampers.

Butterfly dampers are the type of volume dampers that are known to most people. These dampers consist of a flat piece of metal that can be turned in the duct work. When in the full open position, the only resistance caused by the damper is the amount generated by the thickness of the metal making up the damper. In a closed position, the butterfly damper blocks the duct work entirely. Most butterfly dampers allow settings at intervals between the full-open and the closed positions.

Slide dampers are a piece of metal that can be pushed across the opening of a duct. With these dampers, they are either open fully or closed completely.

A hit-or-miss damper is not complicated. This type of duct control is made from two pieces of metal. When a hit-or-miss damper is open, half of the duct interior is unobstructed. If the damper is closed, the duct is blocked off completely.

Any of these dampers should be installed some distance away from the supply air outlet. Otherwise, air flow might be disrupted in an undesirable way.

Splitter dampers

Splitter dampers perform functions very different from the dampers just discussed. Rather than block air flow, like the dampers above, splitter dampers direct air flow. These dampers improve the efficiency of air moving through offsets in duct work. Instead of having a rush of air hit a dead end and bounce in multiple directions at the back of a sharp turn, the air can be guided in a smooth transition with the use of splitter dampers.

Air escaping

Air escaping from duct work can have a dramatic affect on the efficiency of air distribution and the effectiveness of the heating or cooling system. Gaps between ducts, holes in ducts, and other products of poor workmanship or age can cut the efficiency of duct work by 25% or more.

Duct work that is being installed in new construction will not normally be installed in such a way that air leakage is a problem. If the job will be inspected by a code-enforcement officer, workmanship should be of a high enough quality to eliminate most air leakage.

When existing duct work in a home is used as part of a new heat-pump system, air leaks can be a problem, and they can account for a substantial loss in the performance of the heat pump. If you plan to use existing duct work, examine it carefully, and make sure that all seams and joints are well sealed.

Furniture placement

Furniture placement can affect the air distribution from a heating or cooling system. Placing a sofa over a floor register restricts air movement considerably. The same problem can arise if a bed or another piece of furniture blocks the natural dispersement of air from a register. In the case of wall registers, desks and chest-of-drawers can block air movement. Whether you are building a new home or doing a heat-pump conversion on an existing home, take furniture placement into consideration when planning the air-distribution system.

Ceiling fans

Ceiling fans can help to make heat pumps more efficient in some styles of homes. Houses that have tall ceilings are more costly to heat than houses with standard ceilings. Heat rises, so more heat must be produced to keep chalets, A-frames, and others styles of homes with high ceilings warm. One way to improve the heating of such homes is the use of ceiling fans.

The first home I ever built for myself was an A-frame. It was situated atop a mountain, and the peak of the interior ceiling was 24 feet above the floor. The ceiling was made of 2 × 6 tongue-and-groove planks. Because all of the beams were exposed, there was no place to put insulation. The only roof insulation was rigid, board insulation. It was between the ceiling planks and the shingles. This did not make for an energy-efficient home, but the tall, open ceiling and loft had a lot of charm.

The problem I had was one of the heat rising and escaping through the roof. To solve the problem, I installed ceiling fans with reversible motors. The reversible motors allowed me to control the direction of air flow from the fans. I could push heat that was rising back down into the home, or I could pull unwanted heat to the ceiling in the summertime. My heating costs dropped considerably after installing the ceiling fans.

I also used a ceiling fan in the third home I built. This was hanging from a vaulted ceiling in the living room. The house was a unique design, created by my wife, that had two rooms and a bathroom upstairs, with an open, balcony-like hallway. The ceiling fan allowed us to direct and control air flow from the heat pump very effectively. Depending on the design of your home, a ceiling fan might be very beneficial in directing your air flow.

Cold air

Cold air from a heat pump is more important to some users than warm air. For these users, registers mounted in ceiling locations are more effective than floor registers. If you live in an area where air conditioning is more of a priority than heating, you will have to look at your air distribution differently from the way people in colder locations would.

Cold air is more difficult to move than warm air is. This means that you must have larger ducts, more powerful blowers, and so on. If you are converting from a forced hot-air furnace to a heat pump in an existing house, you should plan on installing larger duct work. Dampers can be used to reduce the air volume in the ducts for the heating season, if needed.

When you plan the design and air distribution requirements for your heat pump, you should identify which function of the equipment will be of the most importance to you. In a state like Vermont, the heating ability of a heat pump would be of the most importance. However, a home in south Texas would need superior air conditioning. It is possible to get good service for both heating and cooling from a heat pump, but you should decide which of the functions is the most critical. This will allow you to make minor concessions to customize your heat-pump system.

Hot or cold

Whether you like it hot or cold, or someplace in between, air flow will play a major role in the comfort of your home and the efficiency of your heat pump. Give some serious thought to your personal needs and desires.

10

Heat-pump pads and their proper preparation

Heat-pump pads and their proper preparation is an important part of any successful two-piece heat-pump installation. The outdoor unit of a heat pump must be supported on a firm, level surface. In addition to these requirements, the placement of the equipment pad can have a direct affect on the efficiency of the heating unit. Factor in the many possible locations for the pad, and you've got what could be a confusing situation. This chapter gives you the information needed to make a wise choice on the location of your heat-pump pad, as well as details on how to install it.

Rooftop locations

As discussed earlier, rooftop locations can be used for the installation of heat pumps, but a roof is not the best place to put your equipment. While many commercial buildings use rooftop installations, residential applications for rooftop units are few and far between. This is for two obvious reasons. Most houses don't have flat roofs, and heat pumps that are roof mounted do not function at their best efficiency levels.

Heat pump units that are set on top of roofs are exposed to direct, hot sunshine in the summer. This makes cooling the home more difficult for the heat pump. In winter, the rooftop unit is exposed to brisk winds that handicap the unit's ability to produce comfortable warmth in the home.

While you might see heat pumps set atop the roofs of garages or homes, avoid these locations for your heat pump. Setting the unit near the ground will produce more favorable results all through the year.

Ground-level locations

Ground-level locations are the most desirable sites for installing heat-pump pads. A unit that is mounted close to the ground will be affected less by winter winds than a unit mounted at a higher elevation. With careful placement, a ground-level unit can be shaded by deciduous trees in the summer to improve air-conditioning efficiency. In winter, the trees will lose their leaves and allow the sun to help warm the outside unit. An ideal location will find the heating unit placed with a southerly exposure, that is not close to a bedroom, and surrounded by deciduous trees and shrubs.

Now that you know where the best location for your outside heating unit is, let's see what you must do to prepare for its installation.

Distances

The distances for clearance on a heat pump's outside unit must always be taken into consideration when planning the placement of the equipment pad. Always refer to the manufacturer's specific recommendations for the unit you will be using. Normally, the outside compressor unit will sit at least 30 inches away from the side of the house.

The distance from the air intake on the outside unit and any obstruction should ideally be about three feet. This is not a mandatory measurement, but it is a good practice to observe.

If you are adding landscaping to hide the outside unit from view, keep in mind the growth pattern of your selected shrubs. Many people plant little bushes close to the outside unit only to find that in a few years, the shrubs are blocking the air intake.

The outside unit should be located so that the distance between the inside unit and the outside unit is not more than about 60 feet when using an air-to-air heat pump.

Snow

Snow is another consideration when thinking in terms of distance. If your house is located in a climate where snow is an issue, you must plan vertical distances as well as horizontal distances.

Homeowners with houses in southern states don't have to worry much about deep snow, but residents of northern states do. For example, when I lived in Virginia, my outside unit sat on a concrete pad that was placed on the ground. If that same pad installation had been used for the house I live in here in Maine, the whole unit would be buried under snow right now, and that just wouldn't do.

As I write this, there is about 27 inches of uniform snow on the ground here in Maine, and there are still a few more months of snowy weather to come. The constant cold temperature doesn't allow the snow in Maine to melt very much, so with every new storm, we have increased depth. This type of snow must be planned for when installing the outside unit of a heat pump.

There are many ways to raise the level of an outside platform. Most contractors have a stand made out of angle iron for the unit to sit on. There are some factory-made stands available as well. Standard cinder blocks can even be used to increase the distance between the ground and the bottom of the outside unit. The key is to keep the unit above any anticipated snow build up.

Ground preparation

Ground preparation for the heat-pump pad is very important. You don't want the pad to sink or tip. This can adversely affect the operation of your heat pump. How you prepare the ground to accept your pad will depend on several factors. For example, preparation techniques used when the ground is stable differ from those used when the earth has been disturbed recently, such as in the case of new construction. Another factor in your prep work will be the type of soils that must be prepared. A hard-packed clay will not require as much work as soggy soil will.

New construction

New construction creates a need for more ground preparation than what is normally required for existing homes. Sometimes a spot can be found where the ground has not been disturbed during construction, but often times the areas surrounding a house have been dug and backfilled, creating weak soil conditions.

Installing a pad on soil that has been recently filled in around a house, without the proper preparation, can result in a sinking and tipping base for the outside unit of a heat pump. This is not only undesirable, it can cause the heat pump to malfunction.

To overcome weak soil conditions, spend some time compressing the ground. Contractors often use power-driven tamping machines for this purpose. The tamper can be operated by one person, and it usually runs on gasoline. If you would like to rent such a tamper, you should be able to find one at any major tool-rental center. Be careful, however, not to let the tamper run over your feet. It is a wise idea to wear steel-toe boots when you are running the machine.

When ground is compacted, it should be done in thin layers. This might mean digging a hole in the new dirt at the location for the pad. The hole should be at least 12 to 18 inches deep when you begin the compaction. It is best if you go all the way down to undisturbed ground to begin your work, but this is not always feasible or needed.

Start by running the tamper over the bottom of your hole. When that ground is tight, shovel in about six inches of new dirt. Tamp that layer, and add another layer of fresh dirt. With every six inches of dirt you add, tamp the ground until it is firm. You might find it necessary to wet the dirt with water during the tamping process to achieve the desired results.

When you have a hole that is about eight inches deep, pour in a four-inch layer of sand and compact it. Then, add four inches of crushed stone. Not only will the stone and sand make a firm base, they will allow faster absorption of any runoff water the outside unit produces during the defrosting cycle. Once you have a good, solid base, you can install your heat-pump pad.

Conversion work

Because conversion work is done on existing houses with established grounds, you should not have to go through the tamping exercises described for new construction. When the ground is solid, you can simply dig a hole and install the sand and stone layers as described earlier. A power-driven tamper is not needed under these conditions. You can use the back of a shovel to pound down the sand and stone.

When you are preparing an area for a support pad, the pad should be somewhat larger than the unit it will support. A lot of people make the mistake of digging their holes with outside perimeters equal to the size of their outside units. This doesn't leave any room for the pad to extend past the equipment, as it should. Figure on a four-inch protrusion for the pad around the entire outside unit and your hole should be fine.

Types of pads

There are many types of pads that can be used to support the outside unit of a heat pump. Concrete pads are the most common. These pads can be made on site, or they can be bought from suppliers and brought to the job site for instant installation. Concrete is heavy, but it is one of the best materials you can use as a support for your outside equipment.

Many contractors do little more than place cinder blocks under the outside units they install. This is a cheap and effective method, but it is not nearly as desirable as a good concrete pad. If you opt for a block foundation to support your equipment, make sure you install the blocks in a way that will give uniform, level support.

Foam-filled plastic is another option to consider using as a support for your condenser unit. These pads are lightweight and easy to handle. There are, however, a couple of drawbacks to the foam pads. First, they are expensive. Second, the equipment might vibrate and slide on the pad, unless an antiskid material, such as a sheet of rubber, is placed between the equipment and the pad.

All the houses I've built have been equipped with concrete pads, and I've never had a problem with them. I'm sure cinder blocks and foam-filled pads will do admirable jobs in supporting an outside unit, but I'm a strong believer in concrete pads. You can use any type of pad you like, but be sure the ground foundation for the pad is solid.

Concrete pads

Building your own concrete pad for the outside unit of a heat pump is not a complicated task. You don't even need any special skills to do the job. Assume you have a rectangular condensing unit. Building a pad for this piece of equipment is very simple. Once the ground is prepared properly, you will need a hammer, some nails, a concrete float, a saw, and some lumber.

Measure your heating unit from front to back and from side to side. Add eight inches to those measurements so that the pad will protrude for inches in all directions. Now, cut your lumber to the proper sizes and nail it together to create a rectangle. The size of the lumber you use will depend on how high you want the base to rise above the ground.

Your concrete pad should have a thickness of at least four inches to minimize the risk of cracking; a six-inch depth is better. If you are worried about snow in moderate climates, you can either pour a standard pad and install a snow stand, or you can use wider lumber for your concrete form and have the concrete rise higher off the ground. In real snow country, you should pour a standard pad and use a snow stand.

Once you have the rectangular form nailed together, set it over the compacted earth you prepared. Make sure the top of the form is level in all directions. You might have to trench under one edge of the form a little or bang down on top of it to get all directions level.

Though it is not always necessary, it's a good idea to drive stakes into the ground to hold the form in its desired position. Otherwise, the form might get kicked or moved when pouring the concrete, causing it to slip off the solid foundation you prepared.

Once the form is secure, you're ready for concrete. If you want the pad to have added strength, you can put a layer of mesh wire down for the concrete to set up around.

It doesn't matter whether you mix your own concrete on the site or have ready-mix trucked in, either one will be okay. Pour the first two inches of the concrete on the mesh wire. Then add another layer of wire on top of the wet concrete. Now fill the remainder of the form to the top. This type of construction (with the wire) makes a strong base.

Once the form is full, you must float off the excess concrete. If the top edges of your form are level, you can use anything with a straight edge, such as a 2×4 stud, to remove the overflow of concrete. Pull the stud over the box, and it will skim off any concrete that is above the edge of the form. A concrete float (a flat, metal tool used to finish concrete) is handy for making final adjustments to the finish of the concrete, but really isn't necessary under these conditions.

After the concrete has cured for a day or two, you can knock the form boards off with a hammer. The end result will be a nice, strong concrete pad.

Circular pads

Circular pads are built about the same way that rectangular ones are, except that the concrete form is a little more tricky to build. To build a form for a round condenser, the first thing you will need are some wooden stakes and a roll of metal flashing or garden edging.

Measure the diameter of your outside unit and allow extra room for protrusion when you lay out your form. Drive the wooden stakes into the ground to form a circle of the size desired. The measurement should be taken from inside the stakes, because that is where the metal that will retain the concrete will be placed.

Once you have the stakes forming a circle, unroll the flashing or edging, and place it on the inside of the stakes. Make sure the metal extends all the way to the ground and is level across the top edges. Attach the thin metal to the stakes with nails and you have your circular concrete form. Now you can pour and finish your concrete just as described for a rectangular form.

Why solid ground

Take the time to prepare the ground under your heat-pump pad properly, and provide a solid pad to support your outside unit. Think of the heat-pump pad being to your heat pump what the foundation of your home is to the whole house. If you think along these lines, you will have a trouble-free installation that will last a long time. Here are some of the reasons why the effort is worthwhile.

Tilting

The tilting of an outside condenser can damage your heating equipment. Condenser units use oil for lubrication. If a support pad shifts and allows the condenser to tilt, the lubricating oil might not be able to maintain proper lubrication of the equipment. Would you drive your car without oil in it? No, and neither should you allow circumstances to force your heating unit to attempt to run unlubricated. Saving a little time on the installation of your heat-pump pad could cost you a lot of money later on.

Sinking

Sinking can occur if the ground under an outside unit is not firm. The settling can be a subtle process that doesn't catch a homeowner's attention, especially if the heating unit is camouflaged by shrubbery. As the pad and outside unit sinks, stress will be put on the refrigeration pipes and electrical wires. With enough settling, a broken wire or pipe could result. Again, you are gambling money for repairs when you fail to make a proper pad installation.

Water

Water can build up under a heat-pump pad if the ground around it is not suitable for allowing water to perk down into the earth. That is why you should put stone and sand under the pad. When an outside units defrosts itself, water runs out of it. Without adequate drainage, the runoff water can cause problems.

Snow

Snow that builds up around an outside unit blocks the air intake openings. This causes the heating system to malfunction. Failure to plan for snow by building an elevated pad or using a snow stand can result in a cold house when you need warmth the most.

11

Professional tin-bending techniques

If you've never worked with sheet metal, you might be in for some surprises when you begin to install the duct system for your new heat pump. By planning the design of your ductwork carefully, you can eliminate a majority of the difficult work associated with sheet metal. However, no matter how well your system is laid out in advance, there will be some on-site work with sheet metal required. Even if you were to run flexible ducts for all of your supplies, they are going to have to tie into the plenum.

When someone works with sheet metal, it is generally apparent. There are usually an abundance of cuts and scratches on the individual's hands and arms to prove they have wrestled with the sharp edges of sheet metal. Even professional installers are frequently walking indicators of how dangerous the raw edge of sheet metal can be.

When I entered the trades, my first job was with a mechanical company. The company offered electrical, plumbing, and HVAC services. All of the technicians gathered around the same counter each morning to swap stories and get their job assignments for the day. There was also a fair amount of minor gambling that went on from time to time, usually in the form of a football pool or some other type of pool. One of the favorite topics for the seasoned mechanics to bet on was how many bandages the junior mechanics and helpers would be wearing the next morning.

The shop I worked in used to bend all of their own sheet metal. They had a fabrication shop that was capable of producing all the ductwork required for a job. There are still some HVAC companies that bend their own tin, but many modern companies don't bother with this phase of the work anymore. Instead, they give a take-off list to a fabrication shop and purchase the prefabricated sections of ducts and fittings from the independent fabricator. Unless you have a fully equipped fabrication shop, I suggest you follow their lead and buy as much of your metal as possible from some supplier.

There was a time when all of the ductwork used in heating and air-conditioning systems was metal. This is not the case today. There are coils of flexible duct that see a lot of use in modern duct systems. Some people even use insulated glass fiber board to build their duct systems. If you opt for an insulated board as your trunk line and flexible ducts for your supplies, the work you have to do with metal is kept to a minimum. This idea is well worth considering.

If you have your ductwork made up at a metal shop, you won't have a lot of trouble putting the pieces together. There are, however, some tricks of the trade that can keep the job running smoothly and prevent you from donating blood to your ductwork. This chapter is full of tips and techniques that will make your job of fitting and hanging ductwork easier.

Ordering the metal

Ordering the metal, assuming that you will be using metal ductwork, is the first step in building a duct system. Of course, you will have some planning to do before you can take this step. Other chapters in this book address the design issues that are necessary in planning a successful duct system.

After you have plotted the path of your new ducts, you must create a material list to order from. If you are building a new house, the blueprints might have a mechanical drawing that will show what duct components are needed. In the case of remodeling an existing system, there will be no blueprints to give your supplier. This means you, or someone, will have to create a complete list of all the components needed. The sizes must be detailed and all the fittings should be listed. It is a good idea to include a sketch of the layout with the materials list. The sketch might help the supplier to understand what is needed.

Sometimes suppliers have a good selection of pre-formed metal in stock. At other times, it is necessary for the metal to be fabricated into duct components, and this can take some time. Don't count on running down to the local supply house on a Saturday morning with the intent of picking up your duct supplies to install that same day.

Call around and find out how much lead time is needed to have the ductwork made up.

When you order your metal, discuss with the supplier how the duct will be connected at joints. Drive bars, flat pieces of folded metal, and S-clips are typical devices for creating joint connections. If these devices are going to be used, have the fabricator turn the edges of the ducts to accept this type of connection. It is not unusual for the edges of the duct to just be turned out, in a flange configuration, so that it can be fitted and screwed together. Talk with your supplier to determine how you want to make the connections. Once you have your metal, you are ready to work.

The screws

The screws used to attach ductwork to various components of the duct system should be of a self-tapping variety. When self-tapping screws are combined with an electric screw gun, the results are wondrous. The use of regular screws will slow down the installation process considerably. A pilot hole must be drilled in the duct for each regular screw that is used. The time lost in drilling these starter holes is significant. By simply choosing self-tapping screws, you can save a lot of time. They cost a little more than regular screws, but they are well worth the added expense.

In the old days it was necessary to run the screws into the ductwork with a standard screwdriver. This method will still work, but there are better ways to get the job done. Sheet metal screws have raised, hex-shaped heads. These heads are perfect for using a nut driver on. A nut driver is a device that can be chucked into any standard electric drill. The business end of the driver has a socket that fits over the head of a hex screw. Some nut drivers are magnetized so that the screw will not fall out of the holder. Nut drivers are available at almost any store that sells drill bits. A small investment of only a few dollars will turn your electric drill into an electric screw gun.

Once you are equipped with a drill and a nut driver, screwing your ductwork together will be a breeze. The metal used in ductwork is not very thick, and it allows self-tapping screws to be used easily, and without the use of pilot holes. All you will have to do is put a screw in the nut driver, place the point of the screw on the section of duct you want to penetrate, apply some steady pressure, and pull the trigger on your drill. The screw will dig into the sheet metal quickly.

Hangers

Ductwork is often suspended from floor joists with the use of hangers. There are a number of options available for hanging ducts, but flat strips of metal are the most common type of hanger. These strips can be screwed into the sides of ducts and nailed to the wooden members that support the ducts. Some installers run a band of the metal under the duct and up both sides to a connection point with floor joists. This method provides extra support, but it is not normally needed. Attaching the metal strips to the sides of the ducts will normally be all that is required.

If round ducts, or small rectangular ducts for that matter, are being run in the cavity between floor joists, there are again many possibilities for how the duct will be supported. One fast, easy method is to nail a flat strip of metal across the bottom of the two floor joists that create the duct chase. If a finished ceiling will be installed to hide the duct, the support should be recessed into the joist bay. This can be done by bending the ends of the metal strip and nailing them between the joists. The duct will lay on the metal as a support. Blocks of wood can be nailed between the joists to act as a support, or the duct can be suspended from an overhead hanger that is located between the joists.

Duct tape

Duct tape is used for countless applications. Plumbers use it, carpenters use it, homeowners use it, and HVAC mechanics use it. There are different types of tape used in duct installations, but the old gray standby sees more use than any of the other types. Duct tape is used to seal connections on ducts, to hold insulation in place, to hold makeshift bandages in place on cut fingers, and for many other purposes. There might be occasions when the heat produced in an area warrants a different type of tape. Read the manufacturer's instruction manual that will come with your new heat pump to make sure that gray duct tape is safe to use in the areas you plan to tape. In any event, pick up some duct tape; it will come in handy for lots of jobs.

Getting organized

Getting organized is an important step towards installing your ductwork. Chapter 5 explained what types of tools would be needed on the job. Make sure you have all the proper tools and that they are where you can find them easily. If you will be working in a crawl-space, it is a good idea to use an old, white bed sheet as a command center. Spread the sheet out on the ground and use it as a place to keep all of your tools and small materials. The light in crawlspaces usually leaves something to be desired, and it is easy to lay a tool down in the dirt and have a lot of trouble putting your hands on it again. By keeping your tools on the white sheet, they will be easy to find. Once you have checked over your tools and small materials, such as screws, you are ready to begin installing your ductwork.

Putting the pieces together

Putting the pieces together in your new duct system might require a little help from a second set of hands. Some of the work required when installing a duct system is difficult to do alone. It can be done, but if you don't know the tricks of the trade, getting the job completed can turn your hair gray. The following is a crash course in trade secrets that will enable you to overcome the obstacles of working alone.

There are many different ways of laying out ductwork in acceptable forms. Therefore, I am not going to concentrate on specifics of the system design; however, I will detail how to put the components of the system together. My goal is to get the ductwork installed without bodily injury, and in the quickest, most cost-effective method.

Flex connectors

Flex connectors are used to join duct systems with their equipment in a way that reduces vibration in the ductwork. A flex connector has a collar around the bottom edges of it. This collar can be moved up and down, in an action similar to that observed when an accordion is played. By using a flex connector near the equipment connections, vibration in the ducts is reduced. Without a flex connector, the noise heard through the ducts in the home would be more disturbing. Drive strips and S-clips are normally used to secure flex connectors in place.

Trunk lines

Trunk lines are big and bulky. Working with metal trunk lines alone can be quite a challenge. Trying to suspend these awkward pieces of formed metal from ceiling joists without help can make a person scratch their head. The job is not really that difficult, if you are prepared properly for working alone.

One of the first questions a do-it-yourself might have about the installation of a trunk line involves the connections between the various sections. If the metal has been turned out to accept drive straps, which is a common procedure, the job is so simple that it might seem like you are not doing something right.

Drive straps are flat pieces of metal that have had both edges folded over. The drive straps are open on each end. To use these simple devices, a hammer is needed. Two sections of duct work are butted together. A flange should be turned out on the face of each piece of ductwork. When the two sections are touching, the drive strap is installed by sliding it up the flanges turned out on the ducts. Once the strap is started, it can be driven all the way into place with a hammer. The result is a harmonious connection between the two pieces of ductwork.

Drive straps are not the only way to connect large rectangular sections of ductwork. The fabricator of the duct can create a host of flanges and connection options. For example, each connecting section of ductwork might have an extended tab that is screwed to the section of duct being mated to it. Drive straps are the most common method of connection, but they certainly are not the only way to get the job done.

If you choose to use a glass-fiber board for your trunk line, you will use staples and tape to fabricate your own sections of ductwork. The rigid boards are scored and cut with a sharp knife, in a manner very similar to that used when cutting drywall. The pieces are held in place by staples and tape. When this type of trunk line is butted together, drive straps are not used. The board's rabbeted ends interlock and are stapled and taped.

Rigid insulated duct panels offer several advantages over conventional metal ducts. First of all, they don't slice and dice the installer like metal ducts can. Secondly, the glass-fiber boards are light in weight and easy to work with. Thirdly, the fact that the board is insulated makes it more efficient than metal ductwork.

There are, to be sure, many advantages to working with insulated duct boards. However, because that work is easy and metal work is a bit more complicated, we will continue on our path for installing metal ductwork. Why should we bother discussing metal ductwork if insulated board is so good? Metal is the most common type of duct material used. It is in existing homes and it is installed regularly in new homes. Sheet metal is the traditional material of choice.

I mentioned earlier that hanging a trunk line can be a bit tricky if you are working alone. There are two common scenarios to consider when you are thinking about hanging your own trunk line without help. The first will have you working in a crawlspace. A second situation might find you working in a basement. If your heat pump is in an attic, hanging the trunk line will not be difficult at all, it will basically just be supported by the attic floor. The methods for hanging a trunk line in a crawlspace are quite a bit different than the challenges you might meet in a basement. Because both situations are common, I've included information on them on a one-on-one basis.

Crawlspaces

Installing a trunk line in a crawlspace is, in some ways, easier than doing the same job in a basement. The distance between the earth and the floor joists in a crawl space is usually not more than about two feet. Because the distance between the joists where you are going to hang the ductwork and the ground is minimal, it is easy to prop up sections of duct work as you go. Depending on the height of the crawlspace, you might use bricks, cinderblocks, five-gallon buckets, or some other temporary support to hold your trunk line off the ground while you secure it permanently to the floor joists.

There are drawbacks to working in a crawlspace. Lighting is limited. Space for sitting or standing up is limited or nonexistent. There isn't a lot of room to maneuver ductwork. These bad points are offset, to some extent, by the ease with which the trunk line can be hung.

Basements

Many trunk lines are installed in basements and cellars. These environments generally offer stand-up room for the installer. While it is an advantage to be able to work standing up, juggling large sections of a trunk line in an attempt to hang them from floor joists that are several feet above the ground or floor is another matter.

How you go about providing temporary support for a trunk line will depend, to some extent, on the distance between the bottom of the duct and the floor of the basement or cellar. Unlike working in a crawlspace, you certainly aren't going to prop the duct up on bricks or buckets while you secure it with proper hangers. Instead, you might use stepladders or perforated hanging strap iron.

Two stepladders can be used to hold a section of trunk line in place while you make connections and final hanging arrangements. You can set the duct on the top of the two ladders and work with ease, but suppose the distance between the floor and the joists is more than the ladders can accommodate, what can you do? Perforated strap iron is the answer.

When you are working in a tall basement or cellar alone, get a few rolls of perforated strap iron (the galvanized metal type will work best for this application) and some roofing nails. The strap iron comes in rolls with lengths ranging from 10 to 100 feet, and it is inexpensive. The strap can be cut with any standard aviator snips or tin snips.

Once you know where a section of the trunk line will be installed, cut two pieces of the strapping. Make sure they are long enough to run from one floor joist, under the duct, and back up to a second floor joist. Estimate where the two ends of the duct section will fall along the floor joists. Take each piece of strap iron and nail one end of each piece to a floor joist. You now have two pieces of perforated strap hanging down from the joists.

Gather up your hammer and a couple of roofing nails. Put them within easy reach on your ladder. Take the section of ductwork you wish to hang and move up the ladder with it on your shoulder. Don't worry about getting it in perfect alignment with its point of connection to the next section of duct. Just get it at about the right height and support it with your shoulder. Now you can use your free hand to loop the strap iron under the duct. Bend the strapping up to a second floor joist. The material is heavy enough to stay in place without you having to hold it. Pick up a roofing nail and stick the point of it through one of the holes in the strap and into the floor joist. Now, just pick up the hammer and tap the roofing nail into the wood. Duplicate the procedure for the second strap.

When you are done, your section of ductwork will be hanging from the joists. It will have plenty of freedom for movement to allow you to fit it up with the adjoining section of ductwork. Once you have connected the sections of duct work together and installed proper hangers, you can remove the strap iron.

Once you have the main trunk of ductwork in place, it is time to cut in supply ducts. This really isn't complicated work, but it does intimidate some people. The most difficult part of this process is cutting the metal without getting cut yourself.

Cutting into the trunk line

Cutting into the trunk line can be done in a number of ways. However, before you start to hack away at the sheet metal, you must determine what size hole is needed. You can do this by measuring the take-off fitting that will be installed on the trunk line. One of the easiest ways to do this is to hold the fitting in place and trace around its perimeter with a pencil. The pencil line will indicate an area larger than your desired hole, but it will be simple to measure into the circle or box that you have drawn to allow for the flange of the fitting.

After tracing the fitting, measure the open area of the fitting. You could also just measure from the point of the fitting where you made the tracing to the edge of the inside opening. Either way, measure off of your pencil mark and draw new lines to indicate the open area of the fitting. Now you are ready to start cutting.

There you stand, looking at a piece of metal ductwork, wondering how to get the cut started and what to make the cut with. This part of the job can be done in a lot of ways and with many types of tools. Many professionals begin by taking a screwdriver and driving the point of it through the metal at a position near the center of the opening. Once they have a gash in the metal, they use snips or shears to carve out the opening.

I've seen people cut into trunk lines with reciprocating saws that were equipped with metal-cutting blades, but this creates a lot of vibration. Some people that I've watched work with metal trunk lines begin there holes with a drill bit. You can use whatever method you feel comfortable with to make your starter hole. I suppose drilling a hole in the duct is the neat way to do the job, but, like a lot of professionals, I just gouge a hole into the metal with a strong screwdriver.

Once you have a starter hole opened up, you can cut along the pencil line that indicates the outline of the fitting. Make sure you cut along the inside line that represents the opening of the fitting. If you get momentarily confused and cut along the outside line, your hole will be too large.

What should you use to cut the metal? Snips or electric shears are the best tools for the job. Metal-cutting saw blades will cut the metal, but snips and shears are easier to work with in most circumstances. As you cut the metal, you will notice that it tends to curl. This makes the sharp edges easy to come into contact with, so be careful. Most pros don't wear gloves, but a pair of heavy-duty work gloves can save you from some cuts and scratches.

If you cut enough sheet metal, you will find that there are times when the snips just seem to be marred down in the metal. You might also notice that the metal that has been cut is poised, waiting to strike at your skin. To make cutting the metal easier and to avoid cuts in your skin, trim off excess metal as you go along. The key to cutting the metal without cutting yourself is patience and a strong attention span to what you are doing. If you become rushed or distracted, you are likely to bleed.

Installing the fittings

Installing the fittings on a trunk line is not difficult at all. There are two common ways of making the installation. Some fittings are equipped with several tabs. When the fitting is slid into the hole that was cut in the trunk line, the tabs are bent over to hold the fitting in place. Another method involves fittings that have flanges turned out on them. With these fittings, the flange is placed over the opening in the trunk line and screws are installed through the flange to hold the fitting in place. Neither method is difficult, and both types of fittings work well. Once you have the take-off fittings installed on the trunk line, you are ready to run your supply ducts.

Running supply ducts

Running supply ducts is less difficult than hanging a trunk line. The supplies are much smaller and easier to handle. Depending on the type of ducts you are working with, the supplies might be metal or flexible. A lot of people prefer working with flexible ducts, but metal ducts remain rigid and can be easier to install. Much of the decision between types of ducts will be dictated by the installation requirements.

If you are installing round, metal ducts for your supplies, they will slide into or over the take-off fitting and be held in place with screws. This part of the installation is simple. Just slide the round duct into place and install three or four screws to keep it from pulling loose. When you get to the end of your duct run and want to attach the supply to the register boot, you do it with the same procedure used at the take-off fitting. There really isn't much skill involved for this part of the job.

Flexible ducts are connected a little differently. These ducts look something like overgrown dryer vents, like you see on clothes dryers. They have a corrugated appearance, are usually insulated, and typically attach to fittings with the help of a steel clamp. Basically, you slide the duct over the connection on the fitting and clamp it in place. Again, no big deal.

Returns

The ductwork for cold-air returns will be put together with the same types of fittings and procedures used for other parts of the duct system. The job is no more difficult than working with any of the other ducts.

Avoiding injuries

If you know what to look out for, you can avoid many of the injuries that inexperienced and careless installers incur. Here's a quick run-down of some of the more common accidents that occur when installing ductwork.

Screw guns

Screw guns make putting ductwork together much easier and faster than what would be possible with a standard screwdriver. There are, however, a few potential problems to look out for when working with an electric screw gun. First of all, make sure the tool and its power cord are in good condition and are never placed in water. As I'm sure you know, water and electricity can be a deadly combination.

Electrical shocks are possible with any power tool. While shocks can occur while using a screw gun, there is a more likely type of injury to be aware of. The sheet metal that ductwork is made of is slippery, and it is often round or uneven. This type of material creates a potential risk for having a screw slide off the metal. If you happen to have your hand in the way when the screw slides off, you might wind up with a hole in your hand. It is not un-usual for the tip of a screw to slid on the metal if the screw gun is not held in a way that keeps the screw straight. The slightest angle can result in a painful lesson.

Always keep the point of your screws going straight into the duct-work. Maintain constant pressure on the screw as it is being screwed into the duct. Never put a finger, hand, or other body part in a posi-tion that might be hit by a reckless screw gun. These same rules ap-ply to the use of a drill with a drill bit inserted.

Sharp edges

The sharp edges of ductwork can slice through human flesh quickly and deeply. If you are not careful in the way you handle metal edges, you are sure to get scratched and cut. Some cuts can be pretty seri-ous. The two best ways to avoid getting cut are to wear heavy work gloves and to pay close attention to what you are doing. Many cuts are the result of negligence. If you keep your mind on your work and your gloves on your hands, you can minimize the risk of getting cut.

Puncture wounds

Puncture wounds are not uncommon when people are installing ductwork. If you are using a screwdriver to start a hole in a duct, you must be careful that the bit doesn't slip off the metal and into your body. A flat-bit screwdriver might not seem like much of a weapon, but it can penetrate skin very easily. This can happen when the screwdriver is being hit with a hammer or when it simply slips off a screw and moves under the force of the arm that is using it.

A friend of mine just recently stabbed himself with a dull screwdriver. This man is an experienced mechanic, but he made a mistake. He was holding a clamp with his left hand and tightening the screw of the clamp with his right hand. The screwdriver slipped out of the screw head and embedded itself in his left hand. The bit dug in at the base of one of his fingers, where the finger meets the hand. The wound was very serious. If this can happen to a seasoned professional, it could certainly happen to somebody with less experienced.

Slivers of metal can also cause puncture wounds. When sheet metal is cut with snips, it can develop long, dagger-like pieces of sharp metal. The sharp point and tapered shape of these dangerous pieces of metal can do a lot of damage to human flesh. Stay alert when working around this type of metal and trim off the sharp edges as soon as you can.

Protect your eyes

You should protect your eyes at all times when cutting and drilling sheet metal. The smallest piece of sheet metal can do a lot of damage if it gets into one of your eyes. Buy a good set of safety glasses and wear them. Goggles are often depicted as the best type of eye protection, but most types of goggles tend to fog up and become useless. I've found that safety glasses are much easier to work in.

Secure your ducts

Secure your ducts well when you are hanging them. Even if your hanger is only temporary, make sure that it is strong enough and positioned well enough to keep your ducts from falling. If a piece of sharp duct work falls on you from above, it can cut you to the bone.

Ladders

Ladders can be responsible for some bad injuries. While you will probably only be working with a stepladder during your job, don't underestimate the physical damage that can be incurred from the improper use of the ladder. Use a sturdy ladder, position it solidly, and don't lean out too far on it.

Take your time

When you are fitting ductwork together, take your time. If you get in a hurry or become frustrated with a difficult piece of ductwork, the chance of having an accident increases. Pay attention to your work at all times, and don't try to set new speed records for getting the job done.

Common sense

Common sense is your best friend whenever you are working on a project. If you have basic experience in working with your hands, your common sense can keep you out of a lot of trouble.

12

Understanding returns and supplies

Understanding why returns and supplies are needed in a duct system makes the job of installing such a system a bit easier. The way returns and supplies are installed can have much to do with the performance of a heat-pump system. Some jobs, like conversion jobs, limit the options for installing duct work. In these cases, compromises must be made. However, new construction work allows plenty of freedom for a great duct installation.

There are two basic types of duct systems that might be used effectively with your heat pump. Extended plenum systems are very common in forced-air applications. Perimeter duct systems are also a viable way to deliver forced-air to the many rooms of a home. Either method will get the job done. Within these two basic types of duct systems are many options for modifications. Let's start our discussion with an overview of perimeter duct systems.

Perimeter duct systems

Perimeter duct systems are so named because the duct work extends around the perimeter of the building being served. Supply outlets are typically installed in floors or close to the floors, in exterior walls. Return openings are normally cut into an inside wall, near a ceiling. Warm air is forced out of the supply outlets. As the warm air rises and cools, it is pulled into the return grille and directed back to the heating system. This layout is both efficient, easy to size, and easy to install. There are two common ways to install a perimeter duct system.

Radial perimeter duct systems

Radial perimeter duct systems are simple in design and easy to understand. With this type of duct layout, supply ducts extend from the heating-and-cooling equipment to the supply outlets. Each outlet has its own individual duct that runs from the supply register to the plenum over the equipment unit.

When a radial perimeter system is used with a one-story home, routing the duct work requires little thought. Supply outlets are placed long outside walls, usually under windows, and ducts are run from the heating-and-cooling unit to the outlets. There is no complicated downsizing of ducts to contend with. Each duct is sized based on its length and individual supply need. Then the duct is run, in the most direct manner possible, from the inside unit's plenum to the supply outlets.

When a two-story house is being fitted with a duct system, a radial-system is not always the best choice. The complications of getting individual supply ducts to upstairs locations can create a maze of duct work that is not practical. An extended plenum system can provide less installation hassles in a multistory installation.

Perimeter-loop systems

Perimeter-loop systems, when viewed from the living space of a home, appear to be the same as radial-perimeter systems. The supplies are in the same locations, and so are the returns, but, there is a difference.

A perimeter-loop system, when looked at from beneath the climate-controlled area, will appear very different from a radial-perimeter system. If you look at a radial system, you will see numerous ducts originating at the plenum and extending to individual supply locations. A Loop system will not have as many ducts coming off the plenum, but the ducts that are attached will be larger.

An average perimeter-loop system will have a continuous run of duct work that loops around the entire perimeter of the climate-controlled area. This loop of duct work is fed by ducts that extend from the plenum over the inside unit to the loop. The supply outlets are connected to the loop of duct work.

A typical loop system will have one large loop of duct work and perhaps four supply ducts feeding the loop from the plenum. The return grille for a loop system is installed with the same procedure used for a radial system.

Comparisons

What are the comparisons between a radial-perimeter duct system and a perimeter-loop system? Simplicity is one of the biggest advantages of a radial system. The fact that an individual duct is run from the plenum to each supply outlet makes a radial system easy to understand. It also allows the use of smaller, easier-to-handle duct work, in most cases.

Control over air-flow is easy to manage with a radial system. Closing a damper near the plenum can shut off the air flow to an individual duct with great effectiveness. Because the damper can be located so near the plenum, very little air is wasted in the body of the supply duct.

Loop systems provide uniform airflow, but they can require larger ducts and more expertise in their installation. Making smooth, directional transitions from the feeder ducts to the loop duct is important with a loop system. This is something that is not a factor in a radial system. There might be more duct work involved with a loop system, and this can mean more time and money is involved with the installation.

To use a loop system on the second floor of a two-story home, provisions must be made to hide the duct work making up the loop. Because the duct work cannot be run through ceiling joists, it must hang below them. Under such circumstances, a box would have to be built to conceal the duct work. Not many people would want to have a box built out from the junction of all their walls and ceilings. Even if the box is fashioned in an attractive manner, the cost of it will add to the overall expense of installing such a system. All in all, loop systems are not normally used in two-story homes.

As a homeowner installing a duct system, you will probably have less trouble installing a radial-perimeter duct system than you would putting in a loop system.

Extended plenum systems

Extended plenum systems are very common. They are used in all types of homes and buildings, and they work well, when they are sized properly. Extended plenum systems are so named because a large, rectangular duct extends directly from the plenum over the inside unit to supply air to supply outlets. This extension is frequently called a trunk line.

A trunk line is usually installed near the center of a home, and it runs approximately the entire length of the home. Trunk lines start out big at the inside unit and get smaller as they go along. If the size of the trunk line is not reduced as it is extended, proper air distribution cannot be achieved.

As a trunk line runs the length of a home, supply ducts are cut into it. The trunk line brings air to the supply ducts, and the supply ducts deliver the air to supply outlets. This is similar to a radial-perimeter system, except that the supply ducts are not as long, because they are connecting to the extended plenum (trunk line) rather than running all the way to the inside unit.

Extended-plenum systems can provide better air flow because of their large size. Small ducts create more resistance than large ducts do, and this results in reduced air distribution. Most homes equipped with professionally installed duct work are fitted with extended-plenum systems.

There is another advantage to an extended-plenum system. The branch supply ducts can be installed between floor and ceiling joists, so that they can be concealed with normal ceilings. The building of chases and boxing is needed for the trunk line, but individual supplies can be hidden with relative ease. This is especially desirable in multistory homes.

Duct-free systems

Duct-free systems are not completely free of duct work, but they don't rely on much of it. These systems are not normally used, and when they are used, they are only used for single-story homes built over crawl spaces. Some people refer to the systems as crawl-space plenum systems.

A crawl-space plenum system is unique in that it uses the entire crawl space of a home as one giant plenum. The crawl space is sealed off to be tight and well insulated. Grates are cut into the flooring of rooms above to allow heat to rise from the crawl area. The heating system fills the crawl space with warm air and the home is heated as the hot air rises through the floor and the open grates. This type of system can work for heating systems, but it is not effective for air conditioning.

A crawl-space plenum system would not be used for a heat pump that provides both warm and cool air, but I wanted to mention the system because some people might like the idea of avoiding duct work. If you were to read a book on forced hot-air furnaces and duct work for them, you might very well see crawl-space plenum systems described. The lure of not having to install duct work could be enough to point you in the direction of a duct-free system. Don't do it, at least not if you want your heat pump to keep your home cool, as well as warm.

Insulation

The insulation installed on supplies and returns is important. Without it, the air within the ducts could become too hot or too cold. Not all sections of supplies and returns need to be insulated, but how will you know what to insulate and what to leave bare?

When you install an average heat-pump system, the duct work is intended to convey both hot and cold air to supply outlets. During either of these functions, uninsulated sections of ducts can reduce the efficiency of your system. For example, if you have a supply duct, that is carrying warm air, running through an unheated section of your home, such as an attic, a portion of the warm air will be lost to the space you are not intending to heat. This same duct work will absorb heat from the attic during hot months, when the duct is filled with cool air, causing a loss of efficiency with the air-conditioning system. Let's run down a list of suggestions for insulating duct work.

Any supply duct that is installed in an area that is not climate controlled should be insulated. This could mean insulating ducts in crawl spaces, attics, cellars, garages, attached sheds, and so forth.

Any duct work run on the outside of a home should be insulated. You might think that it is silly to install duct work on the outside of a home, but sometimes it is the only way to get ducts to upper levels. Trunk lines can be run up the sides of homes and enclosed to look like chimneys. This is an effective illusion, but the duct work and the chase-way should both be insulated.

Ducts that develop over long lengths should be insulated. The reason for this is simple. Long duct runs are exposed more than short ducts are. Air outside the duct work can cause the interior forced air to cool or warm excessively. Insulation helps to prevent this unwanted action.

Most people never consider insulating the duct work used in returns. Because returns are bringing used air back to the heating-and-cooling system, it is often assumed that the duct need not be insulated. This is not always true.

Sections of return ducts should be insulated when they pass through areas subject to temperature extremes. If, for instance, you have a section of a return duct passing through your kitchen, it should be insulated. The excess heat generated during cooking can have an impact on the performance of an uninsulated return during the cooling season.

The basics

Without getting too complicated, the rest of this chapter delves into how supplies and returns work, why they work, and why it is important that they be installed in particular ways.

How they work

Supplies supply air to climate-controlled areas. Returns pick up used air and return it to the heating-and-cooling equipment. The principles of operation for supplies and returns are simple.

When working with a heat pump, the air in a supply duct is forced from the inside unit of the heat pump, into the supply ducts, and out of the supply outlets. The blower at the inside unit is responsible for getting the air to the desired locations in the quantity required to provide comfortable temperatures throughout the home.

Once the air has been released into the home, it is pulled back to the inside unit through the return duct. The air is then recycled by the heating-and-cooling equipment.

Why they work

Some do-it-yourselfers are not satisfied by just knowing how supplies and returns work, knowing why they work can be important to dedicated do-it-yourselfers. There is no great mystery to why supplies and returns work. If the ducts are sized and installed properly, they work.

Sizing is critical to the effective operation of a duct system. Ducts that are too large will not produce condensed, forceful air distribution. When ducts are too small, they cannot move enough air to meet the needs of a home. As long as the supply ducts and blowers are sized properly, as discussed in chapter 8, they will provide satisfactory service. The same is true of return ducts.

The importance of proper installation

Some types of duct systems are better suited to certain applications than others. This chapter has explained various types of duct systems. The crawl-space plenum system is a good example of how a duct system that will work for one application is not a choice for another. In the case of the crawl-space system, it could be suitable for heating needs and lacking for air-conditioning requirements.

An extended-plenum system is the type of duct system most often used by professional installers. A perimeter-loop system is not as feasible for two-story homes as a radial-perimeter duct system. Every home has to be evaluated on an individual basis to determine what type of duct system will be best suited to the home. Factors involve accessibility, cost-effectiveness, and overall efficiency.

Installing returns and supplies

Installing returns and supplies in houses being built is no big deal. Finding places to install these same ducts in existing homes is another matter altogether.

One-story homes are the easiest type of home to install duct work in. All of the major duct work is kept below the living space, and it does not need to be concealed with boxes or dropped ceilings.

Houses with multiple stories create a challenge for the installer of duct work. Chapter 6 discussed many construction considerations to take into account when installing a heat pump in an existing home that has multiple stories.

It is important to know how to put all the elements of a duct system together on paper to create an effective supply-and-return system. Assume you have a one-story house that is built over a full basement (the easiest type of job to install).

This installation example examines several installation options. The first job involves the installation of a radial-perimeter system. Assume I have the inside unit set up (chapter 14 shows you how to set up the inside unit) and the plenums are attached with flex-connectors. To make life easier insulated, flexible ducts will be used for the supply ducts. Adjustable round pipe side take-offs will be used in this example to connect the ducts to the plenum. Register boots will be used on the other ends of the supply ducts to direct the flow of air into habitable space.

The return will be made up of rectangular, sheet-metal duct and attached to the return plenum. The fittings needed for this type of installation are minimal. The supply ducts will be installed with sweeping turns to avoid resistance, and the ductwork will be suspended below the floor joists of the home.

Referring back to the last chapter, cut a hole in the plenum for a side take-off fitting to be used with each supply duct. Cut holes in the floor of the home to accommodate the register boots, and attach the boots to the floor. Now you have a point of beginning and ending for the supply ducts. Install flexible duct between the take-off and the register boot. The duct must be supported to prevent sagging and unnecessary resistance. After you have done this with all of the supplies, the only thing left is the return.

All you have to do with the return duct is extend it to a high point on an interior wall that is centrally located in the climate-controlled area. The stack head that mounts on the vertical duct will turn out and accept the mounting of a grille. If you are doing this part of the installation in an existing home, you will probably have to cut open the interior wall, where the return duct is to be installed, to allow for the installation.

You still have your inside unit with the plenum attached to it in the basement, but that is where the similarities end. For an extended-plenum system, start off from the plenum with a starting collar.

You must cut a hole in the plenum and attach the starting collar. With that done, you are ready to extend the trunk line. The first section of the trunk line will connect to the starting collar. Each subsequent section of the trunk line will attach to each previous section of trunk line duct. In most cases, the trunk line will run in a straight direction. Sometimes, depending on the location of the inside unit, the trunk line will run in opposite directions, such as when the heating-and-cooling unit is located in the center of a basement. It is also possible, depending on the design of the home and the placement of the inside unit, that the trunk line will run perpendicular to itself, such as in an L-shaped house. However, for our sample house, the heating-and-cooling unit is set at the far end of the basement, and the trunk line will run straight out of the plenum for nearly the entire length of the home.

Once all of the trunk line has been assembled and supported, you are ready to cut in the branch ducts. The branch ducts might be made up of round ducts or rectangular ducts, depending upon the amount of air-flow required.

Cutting the branch ducts into the trunk line will be just like cutting them into the plenum during the installation of a radial perimeter system. However, some of the take-offs might be made with large, rectangular-style elbows. There will also normally be some volume dampers installed in selected branch supplies. The dampers will enable you to balance the air flow and obtain a smooth-running duct system.

The return for this extended-plenum system is installed in the same way described for a radial-perimeter system. It should rise off the inside unit and extend to a high point on an inside wall.

Fittings for duct work

The fittings available for duct work are numerous. There are fittings for every type of need, and they come in various shapes, sizes, and materials. To expand on this, here's a look at the broad categories of fittings you might use in fabricating your supplies and returns.

Plenums

Plenums are basically large, rectangular boxes that sit above the heating-and-cooling unit. There is one plenum used over the return opening and another installed over the supply opening. Depending on design needs, plenums can be tall, rectangular boxes, sweeping elbow-turn boxes, or shallow rectangular boxes. There is plenty of flexibility in the shapes possible for plenums.

Take-offs

Take-offs used for connections to plenums are available in many different configurations. They can be little more than simple collars to expansive pieces of duct work. There are different angles and shapes that can be used as take-offs. Each type of take-off is rated based on its resistance factor. This is converted to equivalent feet of air movement. The types of fittings used are a pivotal point of the duct design.

Trunk line fittings

Trunk line fittings are made up of angles and elbows. Most of the fittings have slight, sweeping turns built into them. For fittings with sharp turns, diffusers are installed inside the fittings to guide the airflow through the fitting.

Other angles and elbows

Other angles and elbows are used for supply and return ducts. Like trunk line fittings, these fittings tend to be made with sweeping turns that will have a minimum impact on the reduction of air flow. Fittings with the most extreme turns are ones that reduce airflow the most. Whenever possible, turns should be kept to a minimum and should have a small degree of turn.

Boots

Boots are fittings used to connect stacks and registers to branch ducts. They are basically odd-shaped elbows, and they are made in many different designs. These are the fittings used to turn a branch supply from a horizontal run to a vertical run.

Stack heads

Stack heads are fittings used to turn ducts out of walls, such as when a grille for a return is needed. The fitting connects to the duct work in a wall and makes the air-flow take a 90-degree turn.

Wall stacks

Wall stacks are sections of duct work that are made in dimensions to fit inside standard walls. Wall stacks are used for both supplies and returns, whenever the duct must be concealed in a wall.

When the time comes to install your supply and return ducts, there will be plenty of material choices to choose from. Careful planning of the route to be taken by the duct work will give you the best possible air distribution. As you lay out the route, take note of the changes in direction that will be needed.

A metal shop or duct supplier can provide you with prefabricated fittings to make your turns with. While some professionals bend their own tin, you will be much better off to buy fittings and trunk sections that are already made up.

If you are unsure of how to design the proper layout for your ducts, ask for professional help. Take a set of blueprints, or a hand-drawn sketch of the basic layout in your home, to a supplier or contractor. Either of these sources will be able to lay out the duct system for you. Once you have a plan laid out, installing the duct work is not very difficult.

13

Outside installation procedures

Outside installation procedures for air-source heat pumps are not as complicated as you might think. In fact, the process is relatively simple if you take your time, follow manufacturer's recommendations, and pay attention to what you are doing. If you get careless, the job can turn sour, but that is the case with most any type of work people do.

This chapter takes you on a tour of the basic steps involved with installing the outside unit of a heat pump. There are some aspects of this job that might be best left to professionals. The electrical work is just one such example. For the most part, the wiring is the only part of the job that might be too dangerous or too complicated for a homeowner to handle.

There are some rules that must be observed before moving ahead. Not all heat pumps are the same. There are many similarities between various brands, but it is always possible that there will be some particular aspect of a certain piece of equipment that requires special consideration. This chapter gives you a good, general understanding of installation procedures. This text should not be relied upon for the actual installation of a heat pump. Each heat pump comes with manufacturer's requirements and recommendations for installation. It is an absolute necessity that you follow the guidelines provided by the manufacturer.

Failure to handle and install a piece of equipment in accordance with the recommendations of a manufacturer can void the warranty on the equipment. Worse yet, it can result in damaged equipment or personal injury.

The first type of outside unit investigated for installation procedures is an air-source heat pump. This is the type of heat pump that
is most commonly used with residential properties.

Chapter 10 talked about the installation of heat-pump pads,
which is the first step in installing the outer portion of a split-system
heat pump. If you have followed the advice in that chapter, you
should have the pad already in place and ready to accept the outside
unit.

Setting the outside unit

Setting the outside unit is the second step in installing this portion of
your heat pump (Fig. 13-1). Take your time when setting the outside
unit. Be careful not to allow the fins on the coil to become damaged.
Rough handling or bumping into protruding objects with the outside
unit can result in damaged fins, thus resulting in a less efficient heat
pump.

Set the outside unit on the pad, and make sure the pad is large
enough to accommodate it. The pad should extend past the equipment
on all sides. Position the unit so that the connection ports are in a convenient location. These connection ports will include provisions for a
vapor line, a liquid line, a thermostat wire, and the electrical cable.

Once the outside unit is set securely on the pad, it can be connected to the inside unit. This involves only a few steps, but they are
important steps.

Drill the passage hole

You now have the outside unit set into position, and you are ready to
drill the passage hole. The hole needed for the tubing and wiring
must be large enough to accommodate the wires, tubing, the insulation on the vapor tubing, and perhaps a protective sleeve.

The passage hole for a heat pump connection is usually drilled
through wood, but it can be made in the foundation of a home. If the
hole is made in the foundation, most mechanical codes require that a
protective sleeve be installed in the hole to protect the wires and the
tubing passing through it. The sleeve can be made with a short piece
of Schedule-40 plastic pipe. This is the type of pipe plumbers use for
drains and vents.

When you are deciding what size hole will be needed, there are
some aspects of the installation to keep in mind. While the refrigeration tubing has a relatively small diameter, the vapor line should be

TWXO18-060C OUTDOOR UNIT
(ALL DIMENSIONS ARE IN INCHES)

Models	A	B	C
TWX018C-A	$34^3/4$	$33^3/4$	$29^1/2$
TWX024C-A	$34^3/4$	$33^3/4$	$29^1/2$
TWX030C-A	$38^7/8$	$33^3/4$	$29^1/2$
TWX036C-A	$43^1/8$	$39^3/4$	$35^1/2$
TWX042C-A	$43^1/8$	$39^3/4$	$35^1/2$
TWX048C-A	$43^1/8$	$39^3/4$	$35^1/2$
TWX060C-A	$50^7/8$	$39^3/4$	$35^1/2$

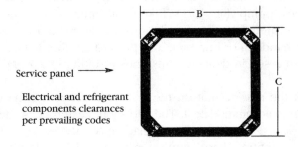

Service panel →

Electrical and refrigerant
components clearances
per prevailing codes

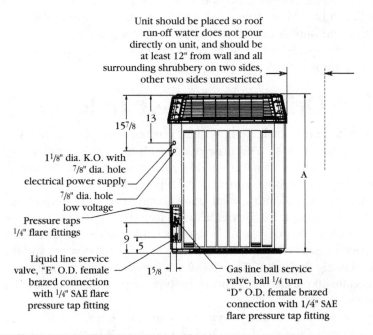

Unit should be placed so roof
run-off water does not pour
directly on unit, and should be
at least 12" from wall and all
surrounding shrubbery on two sides,
other two sides unrestricted

$15^7/8$ 13

$1^1/8$" dia. K.O. with
$7/8$" dia. hole
electrical power supply

$7/8$" dia. hole
low voltage
Pressure taps
$1/4$" flare fittings

9   5

Liquid line service
valve, "E" O.D. female
brazed connection
with $1/4$" SAE flare
pressure tap fitting

$1^5/8$

Gas line ball service
valve, ball $1/4$ turn
"D" O.D. female brazed
connection with 1/4" SAE
flare pressure tap fitting

13-1 *Dimensional data for an outdoor unit.* The Trane Company, an American-Standard
Company

fitted with a protective covering of insulation. The type of insulation most often used is made of a foam-type material and is normally sold in predetermined lengths. It is not uncommon for the vapor tubing to be covered with insulation before it is sold. If it isn't you will have to insulate the tubing yourself. The insulation is applied to the tubing as the tubing is unrolled from the coil it comes in.

The inside of the insulation is coated with a powder-like substance that makes sliding the insulation over the copper tubing easier. There will undoubtedly be a need to seam pieces of insulation together, because the lengths tend to be short. When the insulation is installed and butted together, duct tape can be used to join each new piece of insulation to the previous one.

The insulation creates a bit of bulk on the vapor line, and you must make sure your hole is large enough to accommodate the insulated tubing. This is one step of the installation that many inexperienced installers fail to account for. Consequently, they wind up with holes that are too small.

Typically, a drill bit that cuts a hole with a diameter of two and nine-sixteenths of an inch is adequate. If a protective sleeve is needed, a larger hole might be required. The sleeve is important if the tubing will be passing through a brick, block, or concrete wall. The masonry surfaces are rough, and the copper tubing tends to vibrate. When the copper tubing is vibrated against a rough surface, abrasion is certain, and a leak is inevitable. The smooth inner surface of the protective sleeve keeps unwanted chaffing from occurring.

Most passage holes, however, are drilled through wood. A common location for the hole is in the band board of the house. A band board is the structural member that sits on top of the sill-plate that goes around the foundation. The band board is used to nail the ends of floor joists to. Drilling through the band will put a hole through the exterior wall in a way that the piping can enter an open joist bay (the space between floor joists).

Another option for hole placement is a location that will penetrate the outside wall between the wall studs. This approach might be used if the inside unit was set in a closet that shared the outside wall.

If you drill through wood, you have to penetrate the siding on the home. Once you are through the siding, you will encounter either the band board or a wall cavity. The wall cavity will likely be filled with insulation, and there might be plumbing pipes or electrical wires in the path of your drill bit. This is something you must be careful of.

As the drill bit passes through the siding, it should come into contact with some type of exterior sheathing. The sheathing might be plywood, rigid insulation board, or a composite sheathing. Go slowly with the drill and stop as soon as it is no longer drilling in wood.

If you find insulation board or composite sheathing on the outside wall, break through it with a hammer and a wood chisel. This enables you to inspect the wall cavity for hidden dangers before proceeding with the drill. Once you can see that your way is clear, proceed with the drilling until you have entered the desired location within the house.

Uncoiling the refrigeration tubing

Uncoiling the refrigeration tubing, without damaging it, can be tricky for first-time installers. The tubing will be delivered to the job in a tight coil. Careless handling of the tubing will result in kinks that reduce the effectiveness of the heat pump.

The proper procedure for uncoiling the tubing requires that the tubing be unrolled a little at a time. One effective way to manage this is to use the help of a floor or other surface, such as the ground. Start with one end of the tubing. Turn the coil so that the end of the tubing is on a flat surface. Put one foot, gently, near the end of the tubing. Be careful not to exert much weight on the copper.

With one foot on the tubing, use both hands to unroll a portion of the coil. The copper will be unrolling on the ground and remaining flat and fairly straight. Move your foot to the portion of the unrolled tubing that is closest to the coil. Place the toe of your shoe on the copper and repeat the unrolling process. When you have all of the copper uncoiled, you should be looking at a line of copper that is reasonably straight and unkinked.

Insulating the vapor line

Insulating the vapor line is easy. Simply slide sections of foam-type insulation over the copper you have uncoiled. The insulation should slide on easily. If it doesn't, sprinkle some baby powder into the sections of insulation and shake the insulation to distribute the powder. Powder will make the insulation glide over the copper. After all of the insulation is installed, tape the joints with duct tape. That's all there is to applying the insulation.

Roughing-in the copper tubing

Roughing-in the copper tubing is not hard work, but the job can get tricky. Soft copper tubing bends and kinks easily. This is not something you want to happen. You must handle the tubing carefully and avoid making any sharp turns with it.

Assuming that you are using precharged refrigeration tubing, and you should be, don't cut off the ends. I've had helpers who thought they were planning ahead, doing a good job, and saving mechanics time by cutting the ends off of the tubing. Don't do it. Leave the protective caps on the refrigeration tubing until you are ready to make final connections. Otherwise, the precharged load will be lost, and the copper will no longer be sanitized.

You should begin your rough-in work at the outside unit, but don't connect the tubing to the equipment at this stage. The tubing should be fed through the passage hole from the outside to the inside. Most mechanics tape the thermostat wire to the tubing in order to get it through the hole easily. This can be done with standard duct tape.

Never force the copper tubing if resistance is met. Pushing too hard on the tubing can result in those nasty kinks that you don't want to see. This part of the job is done best with a helper, but it can be done alone. You might, however, have to move back and forth from inside the house to outside the house as you install the tubing.

The complexity of the tubing rough-in will depend, to some extent, on how far the tubing must be run. If you planned your system well, the tubing should not have far to travel. Precharged tubing is available in various lengths. To avoid a situation where you have to cut and couple the tubing, make sure you purchase a roll of adequate length. However, don't get carried away and buy a lot more than you need.

Precharged tubing comes with its own connectors. You don't want to have to cut the tubing, and you don't want to have twenty feet of excess tubing coiled up beside your inside unit. Before buying the tubing, measure the distance between the inside unit and the outside unit. Factor in some extra tubing for any turns and bends that might be needed. Try to buy a roll of tubing that will leave you with as little excess as possible.

The size of the refrigeration tubing to be used for the suction (vapor) line and the liquid line is determined with the use of a sizing chart. The chart normally comes with the heat pump that is purchased.

The two factors that influence the size of the refrigeration tubing are the distance the tubing will be run and the size of the heat pump. Assume that you are looking at the sizing chart that came with your heat pump. To the left of the chart you see a column that refers to the tonnage of the heat pump. For our example, you have a two-ton heat pump, so you locate the two-ton reference in the tonnage column.

The next thing you need to know is how far your tubing will run. For the sake of this example, assume the tubing will extend for a distance of 30 feet, between the inside unit and the outside unit.

As you look across the headings on your sizing chart, you should see provisions for different lengths. The headings will cover a span of distances. For instance, you might see a heading that covers tubing from 22 feet long to 39 feet long. Because your distance is 30 feet, it falls into place under this heading.

Now all you have to do is move down the distance column until you intersect with the tonnage rating. You are likely to see that the suction line (or vapor line as it is sometimes called) should have a diameter of ¾ inch. The liquid line might be rated to be a ⅜-inch tubing.

Sizing charts for refrigeration tubing are easy to understand and simple to use. If you don't have a sizing chart packed with your heat pump, contact the manufacturer for their recommendations.

When you are roughing-in the tubing, you can secure it with temporary hangers. These hangers can be scraps of wire, conventional pipe hangers, or anything else that works for you.

Unless you are fortunate enough to have your inside unit located just on the opposite side of the wall from the outside unit, you are going to have to bend the refrigeration tubing to get it installed. This is not difficult, but it does require some caution, in order to avoid kinks.

The key to bending soft copper without kinking it is to work in small stages, bending a little at a time. Never grab the end of the piping and pull it in a bending motion, you're almost certain of kinking it. Instead, use the palm of your hand as a form for the copper to bend around. Hold the end of the tubing in one hand, and use the heel of your other hand to create the bend. Gently bend the tubing in progressive steps, moving the heel of your hand as necessary.

It is not feasible to make 90-degree turns with refrigeration lines connecting inside and outside heat-pump units. Just as with duct work, long, sweeping bends will give better service. Gentle arcs are what you are after, and with a little practice, they are easy to make.

Electrical wiring

Once you have the refrigeration tubing roughed in between the inside unit and the outside unit, you are ready to run the electrical wiring. This is an area of work that is normally best left to professional, licensed electricians.

Heat pumps require 220-volt electric service. That's a lot of juice to take the wrong way, and an accident when working with such high voltage could be fatal. I'm not a licensed electrician, so I will not begin to tell you how to run the heavy wiring for your heat pump. I will explain the basics of what has to happen, but in no way am I suggesting that you do the work yourself.

The high-voltage wiring for the heat pump will originate at the main electrical service panel in the home. The heat pump will require its own circuit. If the existing electrical service is anything less than a 200-amp service, modifications will be needed. This usually means replacing the existing, undersized, service panel with a 200-amp service. When a complete service upgrade is required, the cost of the work is substantial. Check your electrical service to see that it is a 200-amp service prior to drawing up the bottom line of your installation budget. If money is tight, the cost of upgrading the electrical service could put the job out of reach for awhile.

Assuming that the size of the electrical service is satisfactory, an electrician will run a heavy wire from the service panel to a point outside, near the outside unit. At that point, a disconnect box will be installed. The box should be in sight of the outside unit and easily accessible. A wire will run from the disconnect box to the outside unit to provide electrical power to the equipment.

The low-voltage wiring for the outside unit can be done by do-it-yourselfers who have a reasonable knowledge of controls and wiring principles. Each heat pump will be shipped with instructions for wiring the controls. These instructions should be followed to ensure a proper installation.

Connecting the refrigeration tubing

Connecting the refrigeration tubing is the next step in setting up your outside unit. Before you connect the tubing, check to see that it will not interfere with the service of your outside unit. For example, a misplaced piece of tubing could create an obstacle that would make changing a filter difficult. If you are careless, the tubing might block a path needed by an electrician. Also, make sure that you are allowing enough slack in the tubing to avoid any unwanted contact with walls, joists, or other items.

The liquid line should not be insulated. Only the suction line should be insulated. Check your manufacturer's recommendations pertaining to in-line filters or driers, and follow the directions to install the device if necessary.

When you are ready to connect the tubing to the outside unit, read what the manufacturer's literature has to say about the procedure. Not all precharged tubing is connected in the same way. For the sake of illustrating the type of work you will be doing, here is one common way of connecting the tubing to the outside unit. Keep in mind that your unit and tubing might require some other method.

Most precharged tubing comes fitted with quick-connect couplings. There will probably be female connections at both ends of the tubing. The refrigerant in the tubing is kept in place by a special seal. Once the seal is punctured, the refrigerant is released. You only get one chance to make these connections without losing your refrigerant, so make sure you understand the process before doing anything. Remember to check the advice of your material's manufacturer before proceeding.

When you are ready to make the connection, remove the protective cap from the end of the tubing. It is a good idea to lubricate the threads of the coupling with a few drops of refrigeration oil before attempting to make the connection. The tubing connection that is fitted with an air valve is the one that is normally attached to the outside unit.

Once the threads are oiled, attach the female fitting to the male fitting. Screw the female fitting onto the male threads until it is hand-tight. Be careful not to crossthread the connection. If the female fitting turns freely with the pressure exerted by your fingers, the risk of crossthreading is minimal.

When you get to a certain point in the tightening of the female fitting, you will feel a little resistance. This is due to the automatic puncturing of the inner seal that is holding the refrigerant in the tubing. Once the seal begins to puncture, there is no turning back, at least not without losing the charge of refrigerant.

When you can no longer turn the fitting by hand, use a wrench to tighten the connection. Be careful not to overtighten the fitting. Typically, a half of a turn past the point of a snug fit will be all that is needed. Both the liquid line and the suction line will be attached in this manner.

Securing the tubing

The next phase of the job is to secure the tubing with permanent hangers. There are many types of commercial pipe hangers available. It is also possible to make your own pipe hangers. In either case, the hangers should be used in a way that sound and vibration will not be transmitted to other parts of the structure. This usually means nothing more than making sure there is insulation around the tubing at the point where it rests in a hanger.

The suction line will already be insulated. For this line, hanging it is the only requirement. The liquid line, however, will not be insulated. For this tubing, you must use an isolation hanger or install a short piece of insulation around the tubing where it will come into contact with the hanger.

U-shaped hangers are very common when hanging refrigeration tubing. The hangers are U-shaped with a hole in each ear. If the tubing is being run along a stud wall or floor joists, the U-clamp is placed over the tubing and attached to the wooden members.

Suspended hangers offer a little more protection from vibration and sound travel through the wood. Perforated plastic hanging strap can be used to create hangers of any length. The plastic material allows some freedom of movement and doesn't conduct sound.

The intervals between the tubing hangers will be determined by local code requirements. The code enforcement office might also dictate the types of hangers that are acceptable. You might also find recommendations in the manufacturer's brochures that will guide you in hanging the tubing. Always follow the manufacturer's recommendations and observe local code requirements.

Aside from hooking up the inside unit and testing your work, the process of setting the outside unit is complete. We will talk about connections at the inside unit in the next chapter. Suggestions for testing for leaks and similar installation flaws will also be addressed in that chapter.

Finishing the hook-up procedure

There are some conditions and considerations that might come into play as you are finishing the hook-up procedure for your outside unit. I have saved these considerations and suggestions until now so that they would not confuse the initial steps of installing an outside unit.

Pipe protection

As discussed, sleeves provide pipe protection when refrigeration tubing is passing through a foundation wall; however, that might not be the only type of pipe protection to consider.

If you have your outside unit set two or three feet away from the side of your home, to allow for proper ventilation, there will be some risk that the refrigeration tubing running from the outside unit to the side of the house could be damaged. This risk is greatly increased if you have children.

To a young child, refrigeration tubing can resemble a ladder to be used in climbing to the top of the outside unit. What looks like an ordinary outside unit to you might look like a boat, a fort, or a palace to a five-year-old. If a child uses the refrigeration tubing as a ladder or some other imaginary tool or toy, your new heat pump could be crippled. Soft copper tubing crimps easily, so you can imagine what the tubing might look like after a hard day of playing with the children.

Even if you don't have kids, the refrigeration tubing can be damaged during routine homeowner duties. You might accidentally run a wheelbarrow into the tubing as you are mulching the shrubs in the area. The destructive accident could be as simple as forgetting for a moment that the tubing is there and tripping over it. There are plenty of ways to damage soft copper, so make sure yours is protected.

One simple, relatively inexpensive, and attractive way to provide protection for the tubing involves the installation of lattice. The open weave in the lattice will allow air to flow through to the outside unit while the barrier created will protect the tubing from all except the most deliberate attacks.

Fin protection

Fin protection is another concern that the installation of a lattice barrier can remove. It takes very little to damage the thin fins on the coil of the outside unit. A curious child with a tree branch can do a lot of damage to flimsy fins in a very short time. Because the fins on your coil play an important role in its efficient operation, they should be protected.

The electrical disconnect box

The electrical disconnect box installed near your outside unit might be a hazard to children in your yard. Inquisitive kids could pull the on-off switch to see what happens, or they might attempt to open the box and inspect the wiring. This situation has potential for being very dangerous. A lattice enclosure used to protect refrigeration tubing and coil fins from children can also protect children from electrical connections.

The low-voltage wire

The low-voltage wire that runs from the outside unit to the inside unit is thin and easily damaged. Unlike the high-voltage wiring that is enclosed in armored cable or conduit, the low-voltage wire is protected only by a thin skin of insulation. If this wire is left with a lot of slack in it, the risk of damage is increased. Also, if the wiring is in direct contact with anything abrasive, vibration might eventually wear a hole in the insulation and render the wiring useless. See Fig. 13-2.

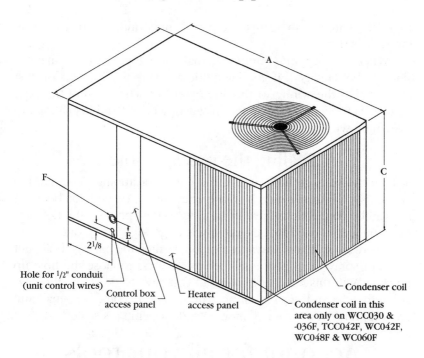

Model	A	B	C	D	E	F
WCC018-024F	55¹/₄	36	25³/₁₆	12¹⁵/₁₆	4³/₈	knockouts for ¹/₂" and 1" conduit
WCC030-036F	55¹/₄	36	29³/₁₆	12¹⁵/₁₆	4⁷/₈	knockouts for ³/₄" and 1¹/₄" conduit
WCC042F	55¹/₄	36	29³/₁₆	12¹⁵/₁₆	4⁷/₈	knockouts for ³/₄" and 1¹/₄" conduit
WCC048-060F	64-⁵/₁₆	45	29³/₁₆	14¹³/₁₆	4⁷/₈	knockouts for ³/₄" and 1¹/₂" conduit

From Dwg 21D729945 Rev. 3

13-2 *Detail of access panels.* The Trane Company, an American-Standard Company

Professional installers normally attach the low-voltage wiring to the insulation surrounding the suction tubing. This can be done with duct tape, but cable ties provide a more positive connection.

Make sure the wire is not trapped against the side of the outside unit and that it is not pinched going into the exterior wall or wall sleeve.

A common mistake made by inexperienced installers is one where the low-voltage wiring is forced into contact with a pipe hanger. The vibration of the tubing that the wire is attached to can cause the wire to rub against the pipe hanger until the insulation on the wire is damaged.

When you support your tubing, make sure the wire is not in a position to be damaged. If you have doubts, wrap a few layers of duct tape around the wiring to provide extra protection within the pipe hanger. You might even want to install a piece of foam insulation between the wire and the hanger.

Sealing the passage hole

Sealing the passage hole, where the wires and tubing penetrate the house, is a job that should be done once the inside unit has been set and all the connections have been tested. Don't plug up the hole before you are satisfied that everything is in good working order.

There are different ways to go about sealing the passage hole, but a can of spray foam is one of the easiest ways to close the hole up quickly. With this insulating foam, all you have to do is aim the nozzle of the can and push a button. The spray will come out small and grow rapidly. In a matter of moments, the opening is sealed.

Account for all your tools

When you have completed the installation of an outside unit, account for all your tools. You would not be the first installer to lose a tool in an outside unit. Having a foreign object, such as a tool, in the outside unit when the start-up is done can create some problems.

Once you have the outside unit set properly, you are ready to move inside. The following illustrations show some of the accessories that are available for heat pumps. (See Figs. 13-3 through 13-8).

13-3 *Cross-section of a heat pump.* The Trane Company, an American-Standard Company

Accesory items – field installed

Part no.	Description	Use with
HPS-UHP	Heat pump stand for "UHP" series	All models
FSM-1D	Fuel saver module	All models
8620-003	Blower time delay relay kit (for "A" coil applications)	All models
8620-060	Outdoor thermostat (0–50°F)	All models
8403-017 8404-009	Thermostat—T874R1129 Sub base—Q674L1181 Manual system change over	All models
8403-018 8404-010	Thermostat T874N1024 Sub base—Q674F1261 Automatic system change over	All models

13-4 *Field-installed accessory items.* Bard Manufacturing Co.

- **Standard Thermostats**

No special thermostats are needed with IMPACK units.

- **Filter Frame Kit**

The IMPACK filter frame accepts standard filters and fits inside the unit. The frame kit functions in either horizontal or downflow duct configurations.

- **Coil Guard Kit**

The guards are vinyl coated 1″ x 3″ wire grills. These grills will protect the coil from hail, kids with sticks as well as normal shipping and installation handling damage.

- **Curb**

One curb fits all the IMPACK models. It ships knocked down. The curb design incorporates the popular locking tabs for quick and easy assembly. Full perimeter curbs available for all models.

- **Economizer**

The economizer fits inside the unit with only the rain hood and barometric relief on the outside. Cabling (with polarized plugs) is shipped with the economizer. This cabling is easily routed to the control box where it terminates in low voltage pigtails. The economizer features a fully modulating low voltage motor eliminating the need for any high voltage wiring. The economizer must be used with the filter frame kit...no return air filter in the economizer kit. A dry bulb sensor is shipped with the economizer. The economizer was not designed for use in horizontal applications. Heat pump applications require a relay kit.

- **Enthalpy Control Kit**

For those applications specifying an economizer with enthalpy control, this control can be used in place of the dry bulb sensor or, alternately, two enthalpy controls can be paired to provide differential enthalpy control.

- **25% Fresh Air Kit**

The kit installs over the horizontal return air opening with six screws for downflow requirements. It can be used on horizontal air flow applications by cutting a hole in the return air duct or in the unit filter access panel.

- **Rectangular to Round Duct Kits**

The adapter kit can be used in either horizontal or downflow applications.

- **Electric Heaters**

One family of electric heaters serves the entire line of 1-1/2 to 5 ton WCC's and TCC's. This will provide the highest degree of flexibility while allowing for minimal inventory level. Outdoor thermoplate kit available.

- **Low Ambient Kit**

An EDC provides low ambient cooling to 0°F with some reduced capacity and protects the system against evaporator icing during other unusual cooling conditions.

- **Fan Delay Relay Kit**

This solid state kit is a time delay that keeps the indoor blower on for about ninety seconds and increases the SEER. It wires into the low voltage unit pigtails.

- **High Static Motor Kit**

Contains a higher torque indoor fan motor.

- **Lifting Lug Kit**

Four reusable lugs in each kit allow units to be easily lifted to rooftop installations. These lugs snap (no screws required) into slots in the unit drip lip channel.

- **Start Kit**

The kit mounts in the control box for those installations specific conditions such as excessive voltage drop due to long wires. (This is a capacitor and start relay kit and is not a PTC device.) This kit can be a good specification buster!

- **Single Power Entry Kit**

The kit minimizes installation costs by reducing the load center circuit requirement and reducing the number of circuit pulls needed.

13-5 *Impack accessories.* The Trane Company, an American-Standard Company

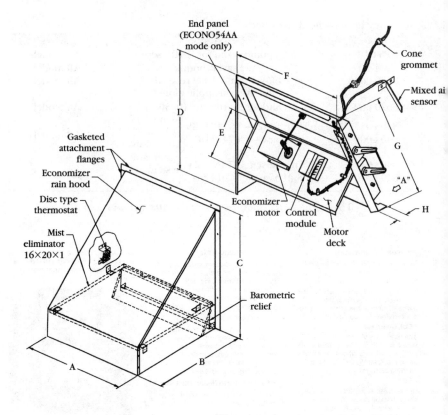

Economizer model	Application models	A	B	C	D	E	F	G	H
BAYECON054AA	WCC018-042F	20	$16^5/_8$	$23^1/_2$	$22^9/_{16}$	$8^5/_8$	$22^1/_4$	$25^1/_8$	$1^1/_2$
BAYECON055AA	WCC048-060F	20	21	26	OMIT	$12^1/_8$	$26^1/_8$	$32^1/_8$	$1^3/_4$

13-6 *Optional equipment for an economizer and rain hood.* The Trane Company, an American-Standard Company

● High Efficiency

IMPACK performance is among the highest in the industry.

● Climatuff™ Compressor

Protection against chemical, electrical, and mechanical stresses are built in for efficiency and a longer life. The compressor is backed by a 5-year limited warranty, with an optional warranty for 5 more years.

● Timed Defrost Control

This is an electronic, time initiated, temperature terminated defrost system that offers a choice of 50, 70, or 90 minute cycles. The time override limits the defrost cycle to 10 minutes.

● Convertibility

IMPACK units are easily converted from horizontal to downflow with the removal of one screw from each panel. Accordingly, the need to stock both dedicated horizontal and dedicated downflow models has been eliminated.

● Installation

The ease of installation and application flexibility exhibited through the design reduce both field time and material.

● Application

The low profile horizontal duct take-offs eliminate the need for expensive transition ducts in crawl space applications.

● Commonality

The common cabinet among the TCC's, WCC's, and YCC's minimizes both the training of sales and service personnel and replacement parts inventory.

● Flexibility

A single curb fits the entire IMPACK line from 1.5 tons through 5 tons thereby providing great installation flexibility on shopping malls, factories, schools, and other commercial buildings where a mix-match of tonnages and utilities is desired.

● Water Integrity

Superior water integrity is accomplished with a water shed base pan having elevated downflow openings and a perimeter channel that prevents water from draining into the ductwork.

● Easy Access

All electrical components can be diagnosed and replaced with the removal of one panel that is attached with two screws.

● Service

All wiring is both numbered and color coded thereby reducing training and servicing costs related to circuit tracing and components replacements.

● Maintenance

A plug on the outdoor fan motor allows the top cover to be removed completely without the hassle of cumbersome wires. The unique service orifice ring allows the indoor fan motor/blower to be removed as a unit.

● Plate Fin Coil

Refrigeration coils are built with internally enhanced copper tubing for high efficiency with less coil area.

● Shipping

Unit dimensions were carefully selected to provide an attractive aspect ratio and for shipping and handling considerations.

● Good Neighbor

Most units can be installed flush with the residence or building thereby minimizing the ground space required. Blankets of insulation reduce blower noise and energy losses to the outside environments.

● Rooftop Mounting

The cabinets are physically smaller than most competitive models. This means less intrusive installations on residential rooftops where aesthetics are critical.

● Handling

The three-way wooden skid allows for easy loading between the wheel wells on pickup trucks for transporting to job sites.

● Structure

The units are lighter weight through the use of high technology components thereby reducing mounting structure requirements and difficulty when man-handling.

● Duct Flanges

Only IMPACK has downflow duct flanges for duct attachments that preserve the built-in water integrity.

● Corrosion

The drain pan is engineered material and eliminates the need for coatings and sealers to prevent sweating and corrosion. The heavy gauge, zinc-coated steel cabinet has a weather resistant enamel finish that stays attractive and protects your investment for years.

● Low Ambient Control

Zero degree ambient cooling is accomplished with two kits. One for low cost installations when full tonnage is not needed. The other kit maintains head pressure and full capacity at zero degrees.

● Quality and Reliability Testing

We perform a 100% coil leak test at the factory. The evaporator and condenser coils are leak tested at 200 psig and pressure tested to 450 psig respectively. In addition the IMPACK designs were rigorously rain tested at the factory to ensure water integrity. Shipping tests are performed to determine packaging requirements. Factory shake and drop tests are used as part of the package design process to help assure that the unit will arrive at the job site in top condition. Additionally, all components are inspected at the point of final assembly. Substandard parts and components are identified and rejected immediately. Every unit receives a 100% run test before leaving the production line to make sure it lives up to rigorous Trane requirements. We at Trane test our designs at our factory and not on our customers!

13-7 *Features and benefits list from a manufacturer.* The Trane Company, an American-Standard Company

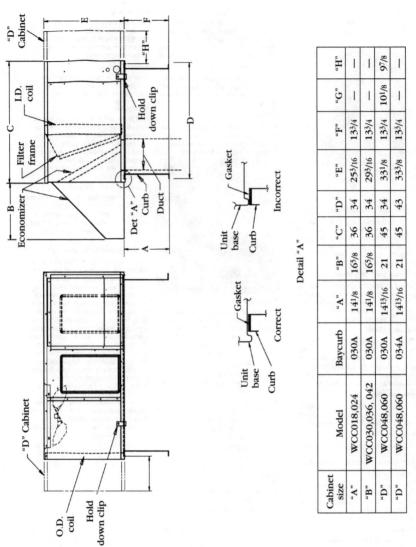

13-8 *Dimensional data for a heat pump.* The Trane Company, an American-Standard Company

Cabinet size	Model	Baycurb	"A"	"B"	"C"	"D"	"E"	"F"	"G"	"H"
"A"	WCC018,024	030A	14¹/8	16⁵/8	36	34	25³/16	13³/4	—	—
"B"	WCC030,036, 042	030A	14¹/8	16⁵/8	36	34	29³/16	13³/4	—	—
"D"	WCC048,060	030A	14¹³/16	21	45	34	33¹/8	13³/4	10¹/8	9⁷/8
"D"	WCC048,060	034A	14¹³/16	21	45	43	33³/8	13³/4	—	—

14

Inside installation procedures

This chapter begins where the last one left off. The heat pump we will be working with is an air-source heat pump. The next step is the inside installation of the heat pump. This chapter also examines installation procedures for water-source and earth-source heat pumps.

Check your list

Before installing an air-source inside unit (Figs. 14-1, through 14-4), there are a few basic suggestions you should observe. While these suggestions are not absolute rules, they will generally enable your heat pump to perform better. Consider it a checklist that should be made before you install your inside unit.

Clearance recommendations

The manufacturers of heat pumps normally include clearance recommendations with the heat pumps they sell. Check the paperwork that comes with your heat pump to see how much clearance is recommended for the inside unit.

A central location

A central location in the home is the best place to install the inside unit. Such a location will allow for less complicated duct work and a heat pump that works better.

WCC018-060F outline–rear
(all dimensions are in inches)

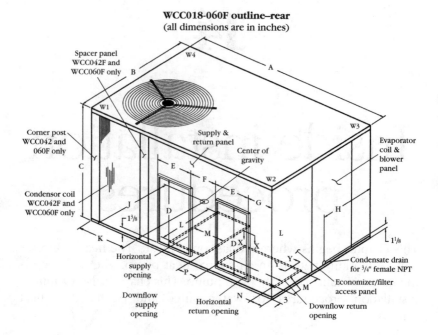

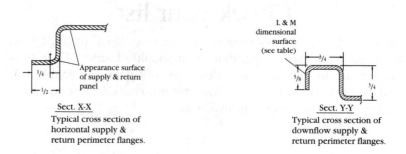

Sect. X-X
Typical cross section of
horizontal supply &
return perimeter flanges.

Sect. Y-Y
Typical cross section of
downflow supply &
return perimeter flanges.

Model	Corner wt. (lbs.)				A	B	C	D	E	F	G	H	J	K	L	M	N	P
	W1	W2	W3	W4														
WCC018F																		
WCC024F	66	55	72	87	55¹/₄	36	25⁵/₁₆	18⁹/₁₆	11¹/₁₆	6⁹/₁₆	6¹³/₁₆	17	20¹/₂	25	17¹/₂	10	3	4⁷/₁₆
WCC030F	77	65	89	105									20¹³/₁₆	25⁵/₁₆				
WCC036F	97	76	85	108	55¹/₄	36	29³/₁₆	18⁹/₁₆	11¹/₁₆	6⁹/₁₆	6¹³/₁₆	17	19	24³/₁₆	17¹/₂	10		
WCC042F	94	73	80	104									18¹⁵/₁₆	24				
WCC048F	126	104	127	153									24³/₄	29				
WCC060F	131	108	132	159	64⁵/₁₆	45	33³/₈	21¹/₁₆	15¹/₁₆	4¹⁵/₁₆	9¹/₈	21¹⁵/₁₆	24³/₄	29	20	14	3¹/₂	8⁵/₁₆

14-1 *Dimensional data and weights of a heat pump.* The Trane Company, an American-Standard Company

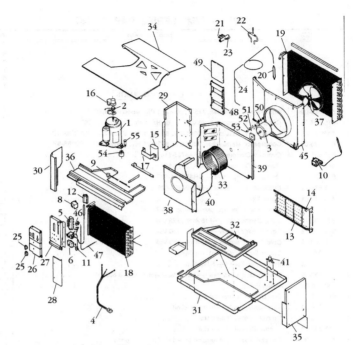

14-2A *Chassis parts.* Friedrich Air Conditioning Co.

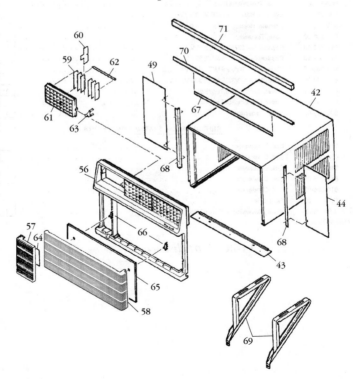

14-2B *Cabinet and mounting parts.* Friedrich Air Conditioning Co.

"YS" — "YM" SERIES PARTS LIST

REF.	PART NO.	DESCRIPTION	YS09H10B	YS09H33B	YS12H33B	YM14H34B
		ELECTRICAL PARTS				
1	615-628-02	Compressor, Tecumseh, 115 V., 60 Hz., 1 Ph., Model RK5490E	1			
1	615-628-03	Compressor, Tecumseh, 230/208 V., 60 Hz., 1 Ph., Model RK5490E		1		
1	01900-420	Compressor, Matsushita, 230/208 V., 60 Hz., 1 Ph., Model 2P19C3R236A-1B			1	
1	615-625-03	Compressor, Matsushita, 230/208 V., 60 Hz., 1 Ph., Model 2K21S3R236A-6B				1
2	603-645-94	Overload, Compressor — MRA3793-114	1			
2	615-780-01	Overload, Compressor — MRA2715-114		1		
2	603-645-77	Overload, Compressor — MRA98781-592			1	
2	603-645-72	Overload, Compressor — MRA98859-628				1
3	610-714-01	Motor, Fan	1			
3	610-714-12	Motor, Fan — for replacement order 610-714-45		1	1	1
4	605-000-13	Cord, Electric Supply — 15 Amp., 125 Volt	1			
4	605-000-18	Cord, Electric Supply — 20 Amp., 250 Volt		1	1	
4	605-000-16	Cord, Electric Supply — 30 Amp., 250 Volt				1
5	606-073-14	Switch, System (Off-On) — 5 Button	1	1	1	1
6	606-072-02	Switch, Fan Speed — 5 Position	1	1	1	1
7	613-503-08	Thermostat (Cool & Heat) — for replacement order 613-503-06	1	1	1	1
8	600-840-31	Relay, Heater		1	1	1
*	616-237-00	Clip, Defrost Thermostat Bulb	1	1	1	1
9	605-668-00	Thermostat, Heater (Klixon), Hi-Limit		1	1	1
10	604-425-00	Control, Defrost Thermostat	1			
10	613-503-09	Control, Defrost Thermostat		1	1	1
11	606-074-00	Resistor, Thermostat Anticipator — 115 Volt, 8,200 ohms	1			
11	606-074-01	Resistor, Thermostat Anticipator — 230 Volt, 33,000 ohms		1	1	1
12	606-122-01	Block, Thermostat Bulb	1	1	1	1
13	605-069-02	Element, Heating — 230 V., 3.3 KW		1	1	
13	605-069-00	Element, Heating — 230 V., 4.0 KW				1
14	606-200-00	Fuse, Thermal		1	1	1
15	610-803-23	Capacitor — 25/10 MFD, 370 V.	1			
15	610-803-10	Capacitor — 25/3.25 MFD, 370 V.		1		
15	610-803-15	Capacitor — 30/3.25 MFD, 370 V.			1	
15	610-803-26	Capacitor — 35/3.25 MFD, 370 V.				1
16	615-422-00	Cover, Compressor Terminal	1	1		
16	614-147-01	Cover, Compressor Terminal			1	1
17	600-627-03	Strap, Capacitor	1	1	1	1
		REFRIGERATION SYSTEM COMPONENTS				
18	616-002-23	Coil, Evaporator	1	1	1	
18	616-002-31	Coil, Evaporator				1
19	615-131-00	Coil, Condenser	1	1		
19	616-005-04	Coil, Condenser			1	
19	616-005-00	Coil, Condenser				1
20	614-813-00	Filter-Drier	1	1	1	1
21	01418892	Valve, Reversing — Ranco V26-109	1	1	1	1

* Not Shown

14-2C *Parts list.* Friedrich Air Conditioning Co.

"YS" — "YM" SERIES PARTS LIST

REF.	PART NO.	DESCRIPTION	YS09H10B	YS09H33B	YS12H33B	YM14H34B
		REFRIGERATION SYSTEM COMPONENTS (Cont.)				
22	603-648-00	Valve, Check	1	1	1	1
23	600-963-00	Solenoid, Reversing Valve — 115 Volt	1			
23	01338857	Solenoid, Reversing Valve — 230 Volt		1	1	1
24	03760520	Capillary Tube — .059 I.D. x 35" Long (Cooling)	1	1		
24	03760452	Capillary Tube — .049 I.D. x 35" Long (Heating)	1	1		
24	03760501	Capillary Tube — .064 I.D. x 32" Long (Cooling)			1	
24	03760482	Capillary Tube — .049 I.D. x 25" Long (Heating)			1	
24	01389975	Capillary Tube — .064 I.D. x 22" Long (Cooling)				1
24	01389903	Capillary Tube — .054 I.D. x 35" Long (Heating)				1
		CHASSIS PARTS				
25	614-939-01	Knob, Control	2	2	2	2
26	604-700-88	Panel, Control (Decorative)	1	1	1	
26	604-700-90	Panel, Control (Decorative)				1
27	614-960-01	Bracket, Control Mounting	1	1	1	
27	614-963-01	Bracket, Control Mounting				1
28	615-826-01	Extension, Control Bracket	1	1	1	
28	615-826-03	Extension, Control Bracket				1
29	615-742-00	Bracket, Control Compartment — Back	1	1	1	
29	615-742-02	Bracket, Control Compartment — Back				1
30	616-141-00	Baffle, Control Compartment	1	1	1	
30	616-141-01	Baffle, Control Compartment				1
31	615-324-15	Base Pan Assembly	1	1	1	1
32	615-300-01	Drain Pan	1	1	1	1
33	606-106-01	Wheel, Blower — 8" OD x 3-⅝" Wide	1	1	1	1
34	615-740-00	Cover & Cross Brace Assembly	1	1	1	1
35	615-733-00	Panel, Right Side	1	1	1	
35	615-733-01	Panel, Right Side				1
36	615-735-00	Deck Assembly	1			
36	615-735-02	Deck Assembly		1	1	
36	615-735-04	Deck Assembly				1
37	605-420-02	Fan Blade	1	1	1	1
38	610-222-01	Front, Blower	1			
38	610-226-09	Front, Blower		1	1	
38	610-226-11	Front, Blower				1
39	615-738-00	Inner Wall Assembly	1	1	1	
39	615-738-01	Inner Wall Assembly				1
40	615-825-01	Wrapper, Blower	1	1	1	
40	615-825-05	Wrapper, Blower				1
41	601-799-00	Valve, Drain Pan	1	1	1	1
42	613-975-08	Shell Assembly (Cartoned with Sill Plate & Logo)	1	1	1	
42	613-975-10	Shell Assembly (Cartoned with Sill Plate & Logo)				1
43	611-098-01	Sill Plate	1	1	1	1

14-2D *Parts list continued*

"YS" — "YM" SERIES PARTS LIST

REF.	PART NO.	DESCRIPTION	YS0910HB	YS0910H33B	YS1212H33B	YM1414H34B
		CHASSIS PARTS (Cont.)				
•	601-691-03	Script for Shell (Friedrich)	1	1	1	1
•	910-029-00	Speednut for Script	2	2	2	2
44	602-944-06	Wingboard	1	1	1	
44	602-944-07	Wingboard				1
45	606-620-03	Shroud, Condenser (Plastic)	1	1	1	
45	606-620-04	Shroud, Condenser (Plastic)				1
46	614-941-01	Slide — Fresh Air & Exhaust Door Control	1	1	1	1
47	615-906-00	Cable — Fresh Air & Exhaust Door Control	1	1	1	
47	615-906-01	Cable — Fresh Air & Exhaust Door Control				1
48	615-840-00	Slide — Fresh Air & Exhaust Door	1	1	1	1
49	615-841-00	Door — Fresh Air & Exhaust	1	1	1	1
50	01237650	Nut — Fan Motor Mount	3	3	3	3
51	606-406-00	Retainer Cup — Fan Motor Mount	3	3	3	3
52	606-405-00	Grommet (Biscuit) — Fan Motor Mount	3	3	3	3
53	01336910	Sleeve — Fan Motor Mount	3	3	3	3
54	610-289-00	Grommet — Compressor Mount	3	3	3	3
55	914-004-00	Screw — Compressor Mount	3	3	3	3
•	614-942-03	Front, Decorative (Complete)	1	1	1	
•	614-942-04	Front, Decorative (Complete)				1
56	614-925-01	Frame, Front Decorative	1	1	1	
56	614-926-01	Frame, Front Decorative				1
57	614-931-01	Door, Control	1	1	1	
57	614-932-01	Door, Control				1
58	614-928-01	Center Section (Intake Grille)	1	1	1	
58	614-929-01	Center Section (Intake Grille)				1
59	614-936-00	Louver, Vertical	18	18	18	18
60	614-936-01	Louver, Vertical Control	3	3	3	3
61	614-934-01	Grille, Discharge Air (Horizontal Louvers)	3	3	3	3
62	610-787-02	Connector, Louver	3	3	3	3
63	614-938-00	Pivot Piece, Discharge Grille	2	2	2	2
64	614-943-01	Label, Operating Instructions	1	1	1	1
65	608-658-06	Filter, Air — 9-½" x 16-½" x ⅜"	1	1	1	
65	608-658-07	Filter, Air — 11-½" x 16-½" x ⅜"				1
66	608-659-00	Holder, Filter	2	2	2	2
•	611-050-00	Accessory Package	1	1	1	
•	611-050-01	Accessory Package				1
•	608-460-05	Hardware Assembly (nuts, bolts, etc.)	1	1	1	1
67	611-097-01	Angle Wingboard (Top)	1	1	1	1
68	611-096-01	Angle, Wingboard (Side)	2	2	2	
68	611-096-03	Angle, Wingboard (Side)				2
69	611-095-03	Bracket, Support	2	2	2	2
70	606-103-02	Gasket, Window Seal (Vinyl)	1	1	1	1
•	01311760	Screw — #8A x ⅝" Hex. Head (for Plastic Front)	2	2	2	2

• Not Shown

14-2E *Parts list continued*

"YS" — "YM" SERIES PARTS LIST

REF.	PART NO.	DESCRIPTION	YS09H10B	YS09H33B	YS12H33B	YM14H34B
		CHASSIS PARTS (Cont.)				
71	600-733-00	Gasket, Window (Foam)	1	1	1	1
•	616-306-02	Carton, Shipping	1	1	1	
•	616-306-01	Carton, Shipping				1
•	616-307-02	Liner, Carton	1	1	1	
•	616-307-01	Liner, Carton				1
•	616-308-01	Pad, Shipping (Top)	1	1	1	1
•	616-246-02	Pad, Shipping (Bottom)	1	1	1	1
•	616-321-02	Pad, Front	1	1	1	
•	616-321-01	Pad, Front				1
		OPTIONAL ACCESSORIES				
•	01900-235	Drain — Condensate Connection Kit, DC-2	x	x	x	x
•	610-089-03	Start Kit, Capacitor/Relay (Pow-R-Pak)	x	x	x	x
•	616-238-01	Winter Cover	x	x	x	
•	616-238-02	Winter Cover				x

• Not Shown

14-2F *Parts list continued*

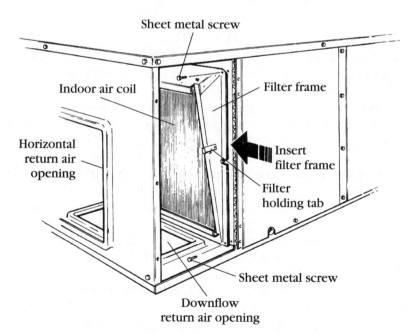

14-3 *Filter frame.* The Trane Company, an American-Standard Company

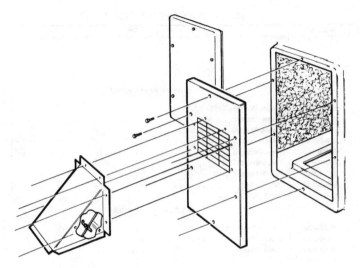

14-4 *Fresh-air kit.* The Trane Company, an American-Standard Company

Close to the outside unit

The inside unit should be installed as close to the outside unit as possible. This reduces the amount of refrigeration tubing required, and it gives the heat pump a little extra edge in performance.

A drain

You'll need a drain in the vicinity of the inside unit. The drain will accept the condensation run-off from the inside unit. This drain is typically a floor drain, but it can be an indirect-waste receptor. An indirect-waste receptor is just a pipe that extends upward to accept the drainage from a hose or pipe. As an example, the drain that a washing machine discharges into is an indirect-waste receptor.

If you don't already have a drain that can be used for the condensate, it might be necessary to call in a plumber. It could also be possible that you can install the drain yourself. If your house is new enough to have plastic drain pipes, you should be able to do the job with minimal effort.

Because many existing homes don't have drains where the inside unit of a new heat pump will be installed, let's take a moment to discuss what is involved with adding such a drain.

If you are installing an inside unit on the first floor of your home, there is little doubt that a drain can be provided for it easily. The plumbing that runs under the home can be tapped into and a drain pipe can be extended to the location of the inside unit.

Assuming that the drainage system in your home is made of Schedule-40 plastic pipe, the job of extending a drain will begin with cutting a new fitting into the existing building drain. To do this, purchase a Wye fitting and hold it next to the building drain. By marking the hub locations of the fitting on the building drain, you can determine how much of the drain pipe must be cut out to accommodate the fitting.

A hacksaw or a regular carpenter's saw cuts the plastic pipe quickly. Try to get a fairly even cut on the pipe. If the end of the pipe is too crooked, the connection at the fitting might leak.

Once a section of the building drain has been cut out, you can install the new fitting. A cleaner/primer and a solvent (glue) is used to make the connections between the fitting and the drain. If there is not enough play in the building drain to allow the Wye to be installed, you must go to plan B. Plan B simply requires that you cut out more of the drain pipe and install the fitting with the help of slip couplings.

Slip couplings slide back on the drain pipe completely and then can be slid onto the pipe extending from the fitting to make the connection. If the building drain won't move in either direction, slip couplings are your best bet for getting the fitting in place.

Once the Wye fitting is cut in, all you have to do is extend plastic piping from the Wye to the location of the inside unit. The pipe should be supported at four-foot intervals, and it should drain from the inside unit to the building drain at a rate of ¼ inch per foot.

Once the pipe is near the inside unit, a trap should be installed on it. From the trap a piece of pipe will extend upward and accept a floor-drain fitting or the indirect waste from the condensate drain of the inside unit. If a vent for the trap is required, and it probably will be, a T fitting can be cut into the drain, near the trap, to extend the vent.

Local plumbing codes will dictate what size piping is used, but the pipe will be no larger than two inches in diameter. Many plumbing codes will also require that the new trap be vented. This can be a problem. Trying to get a vent pipe up through an existing home is not always easy. However, with a one-story home, you should be able to route the vent through a closet and into the attic. At that point the vent pipe can be tied into one of the other vents that penetrates the roof, or the new vent can be put through the roof on its own.

Refrigeration traps

Assuming that your outside unit is set slightly above ground level and your inside unit is set in a basement, you will have to install a trap in the tubing for the suction line (Fig. 14-5).

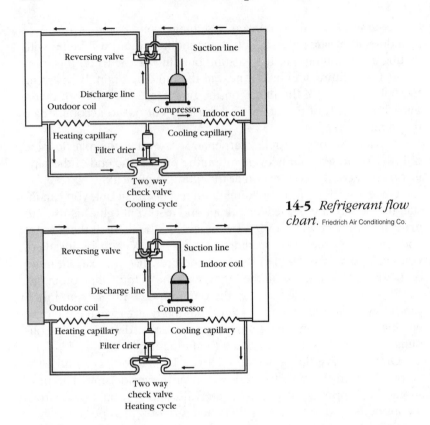

14-5 *Refrigerant flow chart.* Friedrich Air Conditioning Co.

Take your time and bend the suction tubing into the shape of a trap. Let the tubing dip a few inches below the connection port on the inside unit. Gradually bend a loop in the tubing until it lines up with the port in the inside unit. You should be looking at a U-shaped bend, where the bottom of the U is below the pipe connection point at the inside unit. Then the top of the U will sweep over in a gentle bend, towards the inside unit. With the trap formed, you can connect the tubing to the inside unit.

If your inside unit is located in the attic and your outside unit is located near ground level, you will also need a trap, but it will be of a different design. The trap needed will rise above the connection port on the inside unit, rather than dipping below it. This is called an *inverted loop.*

As the suction line from the outside unit is brought to the inside unit, it should bend upward when it gets near the unit. As it rises, it will bend over, in the direction of the inside unit. The loop will run horizontally for a short distance and then turn down. The peak of the loop should be about six inches above the inside unit. As the continuation of the loop drops downward, it should be bent again, to line up with the connection port on the inside unit.

A trap is not needed if the inside unit is installed at a level equal, or nearly equal, to that of the outside unit. For example, if your outside unit is sitting up on a snow stand and the inside unit is on the first floor of the home, a trap will probably not be needed. Under these conditions, you should grade the refrigeration piping so that it falls away from the inside unit and towards the outside unit. If the level of the tubing drops about ½ inch in every ten feet it runs, the system should work fine.

Connecting the refrigeration tubing

Connecting the refrigeration tubing to the inside unit is no more difficult than connecting it to the outside unit was. As with all steps of your installation, it is important to refer to the installation recommendations provided by the manufacturer of the equipment. I've already shown you basically how to connect the refrigeration tubing, so here's a look at what might be a little different about the job than what you encountered with the outside unit.

It is generally recommended that the heater in the compressor be cut for some period of time, up to 24 hours, before the heat pump is fully connected and started. This can be done by providing supply line voltage to the unit. There are different suggestions for ways to clear the units before putting them into operation, so follow the manufacturer's recommendations.

Before you begin to make the tubing connections to the inside unit, take a look around. Will the tubing obstruct the replacement of filters in the unit? Can the electrician work around the tubing without undue trouble? Is there a filter or drier that should be installed during the final connection? Are all areas of the inside unit that might require access for maintenance, such as the blower area, unobstructed by the planned path of your refrigeration tubing? After running through this checklist of questions, you should be ready to mate the tubing to the inside unit. This is generally done in the same manner used on the outside unit, but read the manufacturer's specifications to be sure.

Installing an air-source inside unit

Now that the outside unit has been installed, and the refrigeration tubing has been installed to the inside unit, it's time for the connection. Assume that low-voltage wiring was run to the inside unit, and the high-voltage wiring for the outside unit was in place.

Installing the condensate drain

Installing the condensate drain between the inside unit and the plumbing drain is simple work. The easiest type of condensate-pipe material for most homeowners to use is CPVC plastic pipe. This is a rigid plastic pipe that is sometimes used in making the water distribution systems in homes. The pipe and fittings are joined together with solvent-welded (glued) joints.

The condensate outlet on the inside unit is the place to start your work. Some units might be equipped with male pipe threads while others could be designed with female pipe threads. Obtain a CPVC adapter that will connect to the type of connection present at the inside unit. Pipe joint compound or sealant tape should be applied to the male threads before a connection is made.

Once the pipe sealant is on the male threads, screw the CPVC adapter onto, or into, the connection at the inside unit. Measure the distance from the inside unit to the drain, and cut a piece of piping. CPVC pipe can be cut easily with a hacksaw. The pipe size will typically be either ½ inch or ¾ inch.

It is important to use a cleaner/primer on both the inside of the CPVC fitting and the outside of the pipe that will fit into the fitting. Once the cleaner/primer is applied, the solvent can be swabbed onto the pipe and in the fitting. Insert the pipe into the fitting and turn it at least a half turn. This spreads the solvent and helps to assure a watertight seal.

The joints made with CPVC pipe and fittings take quite a while to set up after being glued. If the joint is jostled, voids might occur in the glue that will cause a leak. To avoid this, have supports in place for the pipe before it is glued into place, and don't wiggle the pipe while working with the opposite end of it.

Depending on the drainage arrangement, you will probably have to install an elbow fitting on the end of the condensate pipe. The elbow will direct the flow into the plumbing drain. Most condensate drains are short and don't require vents or traps, but long condensate pipes and varying code requirements might require either, or both. Check local regulations on this issue.

CPVC pipe is brittle, so it should not be installed in a way that it might be stepped on or hit. Route the pipe to keep it out of the way and protected. More durable types of pipe material can be used to fabricate the condensate drain. If you know how to solder, copper can be used. Other types of materials are available, and you might find that a flexible, hose-type of material will work best for you. Refer to the manufacturer's recommendations and use a material that you feel most comfortable working with.

Electrical work

Electrical work for the inside unit is beyond the capability of most homeowners. The voltage being worked with is high, and so is the risk of injury for untrained people. If you are skilled in electrical work, you will be able to wire the unit by following wiring diagrams provided with it. If you don't have the depth of knowledge needed to work by the supplied diagrams, you should hire a licensed electrician to do the work. I have, however, included some wiring diagrams for various types of heat pumps (Fig. 14-6 through Fig. 14-11). These

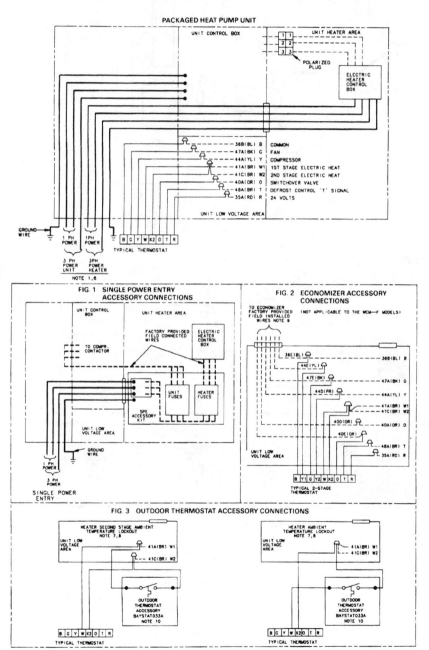

14-6 *Typical field wiring diagram.* The Trane Company, an American-Standard Company

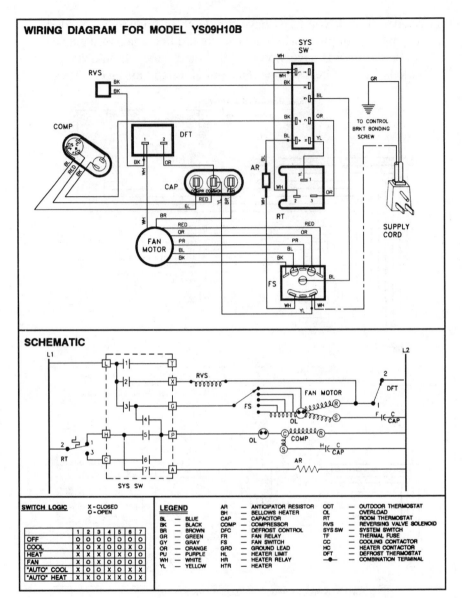

14-7 *Sample.* Friedrich Air Conditioning Co.

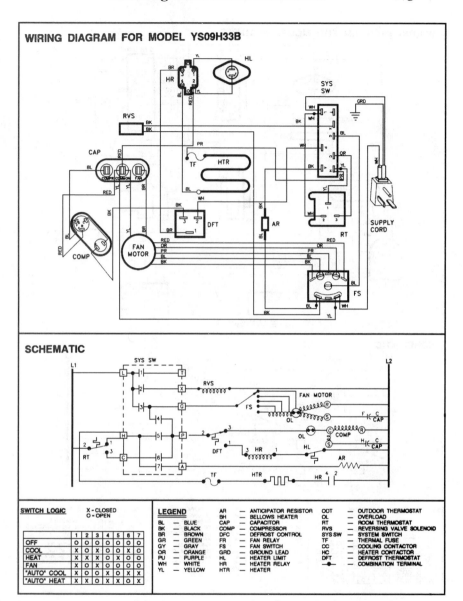

14-8 *Sample.* Friedrich Air Conditioning Co.

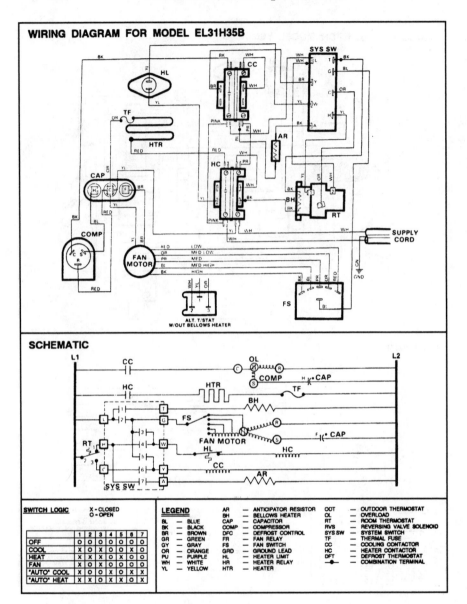

14-9 *Sample* . Friedrich Air Conditioning Co.

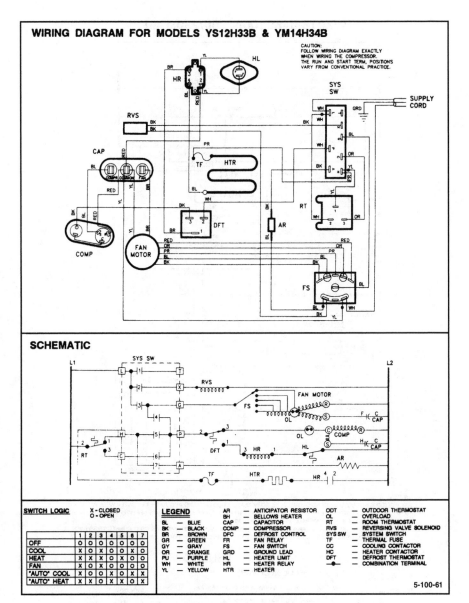

14-10 *Sample.* Friedrich Air Conditioning Co.

TWX018-048C100A: 200/230/1/60

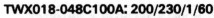

TO 200/230 V., SINGLE (1) PH., 60 HZ. POWER SUPPLY PER LOCAL CODES

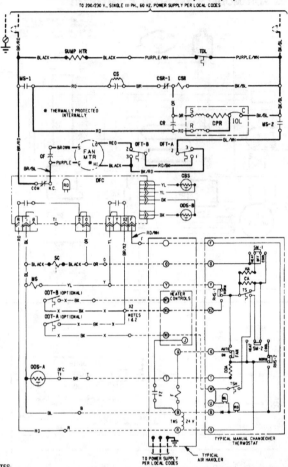

NOTES:
1. IF ODT-B IS NOT USED MAKE NO CONNECTIONS TO BK LEADS AT OD UNIT AND ADD JUMPER BETWEEN W2 & W3 AT A/H
2. IF ODT-A IS NOT USED MAKE NO CONNECTIONS TO BK LEADS AT OD UNIT AND ADD JUMPER BETWEEN W1 & W2 AT A/H.
3. LOW VOLTAGE (24 V.) FIELD WIRING MUST BE 18 A.W.G. MIN.
4. USE COPPER CONDUCTORS. IF ALUMINUM OR COPPER-CLAD ALUMINUM POWER WIRING IS USED, CONNECTORS WHICH MEET ALL APPLICABLE CODES AND ARE ACCEPTABLE TO THE INSPECTION AUTHORITY HAVING JURISDICTION SHALL BE USED.

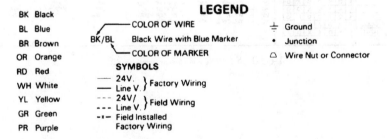

LEGEND

BK	Black
BL	Blue
BR	Brown
OR	Orange
RD	Red
WH	White
YL	Yellow
GR	Green
PR	Purple

COLOR OF WIRE
BK/BL   Black Wire with Blue Marker
COLOR OF MARKER

SYMBOLS

— 24V.
— Line V. } Factory Wiring

--- 24V/
--- Line V. } Field Wiring

-·- Field Installed Factory Wiring

⏚ Ground
• Junction
△ Wire Nut or Connector

14-11 *Sample schematic diagram for wiring.* Friedrich Air Conditioning Co.

TWX060C100A: 200/230/1/60

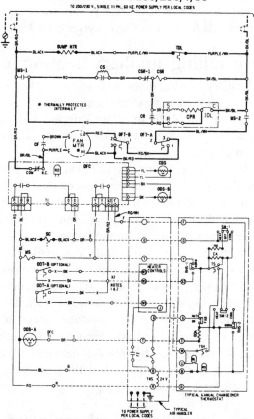

NOTES:

1. IF ODT-B IS NOT USED MAKE NO CONNECTIONS TO BK LEADS AT OD UNIT AND ADD JUMPER BETWEEN W2 & W3 AT A/H.
2. IF ODT-A IS NOT USED MAKE NO CONNECTIONS TO BK LEADS AT OD UNIT AND ADD JUMPER BETWEEN W1 & W2 AT A/H.
3. LOW VOLTAGE (24 V.) FIELD WIRING MUST BE 18 A.W.G. MIN USED, CONNECTORS WHICH MEET ALL
4. USE COPPER CONDUCTORS. IF ALUMINUM OR COPPER-CLAD ALUMINUM POWER WIRING IS
   APPLICABLE CODES AND ARE ACCEPTABLE TO THE INSPECTION AUTHORITY HAVING JURISDICTION SHALL BE USED.

LEGEND

∿ Coil	☐ Terminal Board	CSR Capacitor Switching Relay
⊣⊢ Capacitor	⊸⊩⊸ Diode	DFC Defrost Control
⊣⊢ Relay Contact (N.O.)		HA Heating Anticipator
⊬ Relay Contact (N.C.)	Pol. Plug Female Housing	IOL Internal Overload Protector
	(Male Terminals)	MS Compressor Motor Contactor
⋋ Relay Contact (SPDT)	Pol. Plug Male Housing	ODA Outdoor Anticipator
⊕ Thermistor	(Female Terminals)	OFT Outdoor Fan Thermostat
	⊏⊃ Light	ODS Outdoor Temperature Sensor
ᴓ Internal Overload Protector		ODT Outdoor Thermostat
⌐ᵔ Temp. Actuated Switch	CA Cooling Anticipator	RHS Resistance Heat Switch
∿ Switchover Valve Solenoid	CBS Coil Bottom Sensor	SC Switchover Valve Solenoid
	CF Fan Capacitor	SM System "On-Off" Switch
⋙ Resistor or Heating Element	CN Wire Connector	TDL Discharge Line Thermostat
⌐⌐o Motor Winding	CPR Compressor	TNS Transformer
	CR Run Capacitor	TS Heating-Cooling Thermostat
o Terminal	CS Starting Capacitor	TSH Heating Thermostat

diagrams are only examples to illustrate what you will be likely to see from the manufacturer of your heat pump, they are not intended to be wiring methods for you to use on your own system.

Installing auxiliary heating units

Installing auxiliary heating units is necessary with some types of heat pumps. Almost all heat pumps are equipped with some type of auxiliary heat, but not all of the units are shipped with the heating units installed. The most common type of back-up heat is electric resistance heating elements.

If electric heating elements are not factory installed, they can usually be installed on site quite easily. Most electric heating units simply slide into an allocated slot in the inside unit. It is very likely that the back-up heat will already be in place, but you should refer to the manufacturer's installation recommendations to be sure.

If you are doing your own wiring, you might have to make special provisions for the outdoor thermostat, or thermostats, that control the auxiliary heating units. Some manufacturers ship their heat pumps with the outside thermostat already installed and wired, so refer to the installation guide for the particular model of heat pump you are installing. Also, expect to find a heat relay for units that use multiple outside thermostats. This relay is used to bypass the outside thermostats if the heat pump is not functioning properly and the auxiliary heat must be relied upon as a primary heat source.

Installing the thermostat

The most difficult part of installing the thermostat for the inside unit is fishing the wire from the inside unit to the thermostat. There are, however, some general guidelines that should be followed when selecting a location for the thermostat.

As a rule of thumb, the inside thermostat should be located above the floor at a height equal to chest level for the occupants of the home. In terms of actual measurements, a height of five feet is common.

Thermostats should be mounted on inside walls, and they should not be placed near objects that will affect their performance. Such objects could include supply ducts, direct sunlight, heat-producing appliances, drapes concealing the thermostat, and so on.

Testing for leaks

Testing for leaks in the system is the next step towards completing the job. Simple testing of the refrigeration connections can be done with a soap-and-water solution. Use the soapy mixture by applying it around the finished connections. If there is a leak, the solution will begin to bubble. Tightening the connection will probably stop the leak and the bubbling.

Professional test methods are considerably more complicated than the bubble test. To test a heat-pump system properly, special test gear and knowledge is needed. This is a good job for professionals to be called in for. Because such testing is beyond the capabilities of most homeowners, I won't spend a lot of time on the procedure, but I will hit the high spots for those who are curious.

Professional heat-pump mechanics use gauges, leak-detecting probes, and nitrogen or refrigerant to test systems for leaks. A tank of refrigerant or nitrogen is attached to a gauge manifold. The cut-offs on the manifold are opened and the contents of the tank are allowed to pressurize the heat-pump system.

Once the system is pressurized, professional installers normally use electronic leak detectors to check the system out. The probe is passed over, below, and around possible leak locations. If a leak is discovered, its location is noted and the testing is continued.

When the entire system has been checked out, any leaks that were located are fixed. This requires draining the system and brazing the connections. Then the test is reapplied to make sure the leaks were fixed.

Purging

Purging a new system is not normally required as long as precharged refrigeration tubing was used during the installation. If the tubing used for the job was not precharged, purging will be necessary. Also, if the inside coil is allowed exposure to the atmosphere for more than just a few minutes, purging of the system should be done.

Because purging is not normally needed with the installation of a new heat pump, I won't go into details on how the procedure is performed in this section of the book. Specifics on purging is included later when troubleshooting and correcting problems is addressed.

Putting an air-source inside unit into operation

Putting an air-source inside unit into operation is a job that is often best left to professionals (Table 14-1). This is not to say that a homeowner can't get the job done, but due to the technical work involved, professionals are often the best solution.

Table 14-1 Pre-start-up checklist for air-source systems

Phase	Checked
Voltage	
Wiring	
Auxiliary heat is installed	
Refrigerant lines are not crimped	
Refrigerant charge is correct	
Condensate line is unobstructed	
Blower turns freely & speed is correct	
Filters are clean and installed	
Packing materials are removed	
Access panels are in place	
Thermostat is turned off	

If you decide to see your job all the way through, without expert assistance, be sure to read and follow the manufacturer's recommendations. There are many different types of controls in use with modern heat pumps, and it is mandatory that specific start-up instructions, for the unit you are installing, are followed.

While I can't give you guaranteed start-up instructions for your heat pump, I can tell you how many heat pumps get put into operation. The first step is to check all aspects of the installation to see that they are in accordance with the manufacturer's recommendations. Assuming that they are, you can test the limit switch. To do this, you let the heat pump come up to its working temperature with the blower turned off.

The crankcase heater should be turned on well in advance of the start-up procedure. How far in advance? That depends on the manufacturer's recommendations, but the time will normally amount to several hours. All electrical work should be tested to see that it is the proper voltage and amperage.

When the heat pump is started, let it run for awhile before you begin to inspect it. While it is running, listen to the sounds being made. If you hear anything that sounds unusual, such as a straining motor or a clanking sound, check it out. Check the refrigerant charge in accordance with the manufacturer's recommendations.

Once the heat pump has stabilized, check the thermostat to see that it works. If you are operating in the heating mode, wait until the heat pump has cut off, and then turn up the thermostat. The heat pump should kick back on. If you are testing in the air-conditioning mode, you would turn the thermostat to a lower setting to cause the heat pump to turn on.

Refer to the manufacturer's recommendations, and put the heat pump into a manual defrost setting. Let the unit run through the defrosting cycle as you monitor its actions.

With all of the above steps done, you can go from room to room and check air distribution. It might be necessary to adjust damper controls to balance the system and achieve the air flow desired.

Water-source systems

While most homeowners will be installing air-source systems, some people will find other types of systems to be more beneficial to their needs. The installation of water-source systems requires work that is not involved with air-source heat pumps. The fact that water is used as a heat source makes the installation requirements considerably different. The paragraphs that follow will give you a good overview of how water-source heat pumps are typically installed, but don't rely on the information to pertain to the particular heat pump you will be installing. It is always best to read and follow the manufacturer's recommendations when installing any type of heat pump.

Most residential applications of water-source heat pumps involve the use of vertical units. It is, however, possible to buy units that are meant to be installed horizontally. Horizontal units might be worth a look if you are restricted in available space for installing a vertical system. Because most residential water-source heat pumps are vertical models, that is the type discussed here.

The heat pump unit can be installed on any flat, solid surface. Before the unit is set in place permanently, a sound insulator should be installed beneath it. Such insulating material is most often made of rubber. The sound insulator should be installed on top of the floor or mounting platform and under the heat pump.

Electrical hook-ups

Electrical hook-ups for the unit will include a disconnect switch that is installed close to the unit. The disconnect box is normally installed within three feet of the equipment, and it should be in an easy-to-see location. The wiring for the unit is usually protected by conduit or armored cable, and it must be properly grounded. All electrical work must, of course, meet local code requirements.

Piping procedures

Piping procedures for water-source heat pumps vary. There are many possibilities for possible piping configurations, so we will only discuss some general specifications. Refer to the installation instructions that are packed with your new heat pump to make sure you provide the proper piping design.

Piping procedures, as discussed here, include more than just the installation of pipe. There are a number of accessory items that are often used with water-source heat pumps. These items are installed in conjunction with the water piping, and therefore are considered a part of the piping procedures.

Pipe size

Pipe size is one of the first issues you will have to deal with when installing piping for a water-source heat pump. The size of the pipe will be determined by manufacturer's recommendations and the gallons of water per minute that are needed to make your particular system function properly. A quick look at the paperwork accompanying your new heat pump should provide the information needed to select the proper pipe size and type.

A water filter

A water filter is often recommended on the inlet piping. The filter acts as a strainer to trap foreign water-borne particles before they can enter the heat pump. Logically, the filter is installed on the supply piping that feeds the equipment. Of course, the filter is installed before the supply piping enters the heat pump.

The types of filters used on inlet lines vary, but they are typically of an in-line type. This simply means that they are installed on the pipe that is bringing water to the heat pump. The housings for these filters are usually made of plastic, and care must be taken not to melt the housing if soldering nearby is required. It is also important not to crossthread fittings being screwed into the filter housing.

If you are using copper pipe on the inlet line, you should use union fittings at the connection point with the filter. A threaded nipple can be screwed into the filter housing and the union can be screwed onto the other end of the nipple. Most filter manufacturers recommend a pipe-sealant tape be used in place of a standard pipe-thread compound for these connections.

Never thread in a copper fitting and solder a joint between the threaded adapter and the supply pipe. The heat that transfers down the copper might melt the filter housing. By installing a union, you can avoid any heat damage and make a quick connection to the filter.

When installing an in-line filter, you should install at least one valve on the supply side of the filter. It is better to install a valve on each side of the filter. This allows the filter to be changed with minimum effort.

Solenoid valves

Solenoid valves are often installed on the outlet side of heat-pump piping to maintain a desired water pressure. In addition to solenoid valves, other valves, such as gate valves or ball valves, are often installed to facilitate the easy cleaning of the water coil in the heat pump. The installation instructions and diagrams that are packed with your heat pump should suggest exact installation instructions for these valves.

Automatic valves

Automatic valves that control the flow of water on the outlet side of the heat pump are needed during the installation phase. These valves might be either pressure-activated or controlled by low voltage electrical power. If the expected water temperature will be below 60 degrees, the pressure-activate type is normally recommended.

When a low-voltage valve is used, it is installed on the discharge piping. As the pipe comes out of the heat pump, there should be a ball valve installed. After the ball valve is in place, the low-voltage valve is installed.

Pressure-activated valves are normally installed in pairs. One of the valves works at a low-pressure range, and the other operates at a high-pressure range. When a dual installation of pressure-activated valves is done, the two valves are installed close to each other. Recommended piping arrangements for the valve installation could vary, so read your manufacturer's suggestions closely.

Condensate drain

The condensate drain for a water-source heat pump will be installed using the same methods described for air-source heat pumps.

Flowmeter

A flowmeter should be installed on the inlet piping of the heat pump. The flowmeter allows the flow of water coming into the heat pump to be checked. Typically, the flowmeter will have female threads at each end of it. The device is installed in the inlet piping by using threaded adapters that allow the flowmeter to mate with the piping material being used. It is important to note that the flowmeter should be installed in vertical piping.

Closed-loop systems

Closed-loop systems require special equipment to deal with pumping water, fluid expansion, and air. This special equipment is nothing exotic, but it is needed. There are two ways to go about obtaining and installing the gear. You can go out and purchase each part individually, or you can buy a kit that includes all of the parts. Most people favor buying the kit.

When you buy a pump kit for a closed-loop system, there is a good chance much of the material will already be assembled. Not all kits are the same, but they are all designed to perform the same function.

Flowmeter

One of the parts in the pump kit will be a flowmeter. This device is installed vertically on the incoming pipe from the closed loop.

Air purger

The air purger will be the next component of the kit to be installed. This device is installed horizontally and accepts the installation of a pressure gauge and an air vent. The air vent is a type commonly called a bottle vent. Both the vent and the pressure gauge should be supplied with the pump kit.

Expansion tank

There will be an expansion tank in the pump kit that is installed under the air purger. A thermometer will also be provided for installation at the expansion tank.

Isolation Valve

As the inlet piping comes out of the air purger, it should enter an isolation valve. The valve will be part of the equipment supplied in the pump kit.

Circulating pump

The last piece of the puzzle is a circulating pump. The inlet piping leaves the isolation valve and connects to the pump. From there, the piping enters the heat pump.

Service valves

A service valve might come with the pump kit, but two of them will be needed. One of the valves will be installed on the inlet piping prior to the flowmeter. The other valve will be installed on the outlet piping as it leaves the heat pump.

Location

The location for all of this equipment should be chosen carefully. There must be adequate support in the area to hold the weight of the equipment, and the set-up should be installed as close to the heat pump as possible. Typically, the pump kit is mounted to brackets that are secured to wall studs.

Bringing a closed-loop system online

Bringing a closed-loop system online requires a little more work than what is involved with an air-source heat pump (Table 14-2). The loop must be flushed thoroughly before the system can be put into operation. Both the supply and return piping must be flushed out. To do this, you are going to need a flushing pump, some 1¼" pipe, a watertight container, and some clean water. You might want to have a professional take care of the flushing and testing of the system, but if you decide to do it yourself, the following suggestions will give you a good idea of what you are getting into.

Table 14-2 Pre-start-up checklist for closed-loop systems

Phase	Checked
Voltage	
Wiring	
Water system cleaned and flushed	
Air purged	
Isolation valves open	
Condensate line is unobstructed	
Blower turns freely & speed is correct	
Filters are clean and installed	
Packing materials are removed	
Access panels are in place	
Thermostat is turned off	

The pump used to flush the system must be powerful enough to get the job done. This usually means a pump with a horsepower rating of at least 1½ horsepower. The piping run from the pump to the loop lines should have a large diameter, something in the neighborhood of 1¼ inches. Stainless steel clamps can be used to attach the flexible piping to the pump and the loop lines. The working pressure for the test does not have to be extreme, a pressure of 35 to 40 pounds per square inch should be adequate. The flushing procedure will involve both the inlet and the outlet piping. In other words, you will wind up performing the procedure twice, once for each line.

The watertight container will be used to hold water used during the flushing. The container should be clean and free of debris. The size of the container will depend on the amount of piping that has been installed in the loop. A 55-gallon drum works very well for this procedure, but a smaller container can be used. It might be necessary, however, to refill the container during the flushing cycle, so have a clean water source available. A garden hose hooked to your home's hose bibb will do fine.

Once the pump is connected to the first loop pipe and the pickup pipe for the pump is in the container of water, you are ready to begin flushing the system. To start with, close the isolation valve installed with the pump kit to block any water from entering the heat pump. Your goal is to force a full flow of water through the loop.

Before you start the pump, check the suction pipe in the container of water. The end of the pipe should be completely submerged when the pump is running. It is also important that the suction pipe doesn't pick up debris from the bottom of the barrel. Some mechanics attach strainers to the suction line to reduce any likelihood of sucking up foreign particles. The strainer could be a foot valve, like those used on regular water pumps, a piece of screen wire, or even a section of ladies' hosiery. As long as the mesh of the strainer is open enough to allow a strong flow of water and small enough to block out debris, the material used is not important.

A return line will need to be installed temporarily for the water being used during the flushing to be returned to the storage container. This return line can be connected to one of the boiler drains installed on the lines. The end of the return line should also remain below the upper water level at all times.

When everything is set, start the pump. You should see the water level in the barrel dropping until it starts to be replenished from the return line. If the water level begins to drop to a point near the end of the suction or return pipe, stop the pump, and add more water to the container.

Once the return lines begin to replenish the water supply, the flushing should continue without the need for additional water. The length of time spent flushing the system will vary, but expect to spend at least 15 minutes on both the inlet and outlet piping. Basically, you should continue the process until the water coming out of the return pipe no longer contains air bubbles. If the water container begins to fill with debris during the flushing cycle, the pump should be stopped and the water in the container should be replaced with clean water.

When the flushing of the first loop line is complete, close the service valves and cut off the pump. Connect the pump to the second loop line and repeat the procedure described above. Once the second loop line is flushed thoroughly, you are going to flush out the coil of the heat pump. This is done by opening the isolation valve on the pump kit, while the flushing pump is still pumping. You should see activity in the flowmeter at this point. It is best to turn on the circulating pump that was installed with the pump kit to ensure a full flow of water through the system. When there is no more air or debris being pumped into the storage container, you can stop the flushing operation.

Adding antifreeze

The next job to be tackled is adding antifreeze to the loop. This is not a difficult procedure. In fact, it is very similar to the work done when flushing the system. The antifreeze used will likely be either calcium chloride or proprolene glycol. Calcium chloride is less expensive than proprolene glycol, and it is probably the more popular type of antifreeze for this reason.

Before you can charge the system with antifreeze, you must know how much should be added. This can be determined with volume charts that are typically supplied by heat pump manufacturer's. Assuming you have a chart, you need only to know how many feet of pipe were used in the construction of the loop, the type of pipe used, and the diameter of the pipe. For example, let's say you installed 400 feet of 1¼" polybutylene pipe in your loop. By looking at the chart provided with your unit, you might see that 7.8 gallons of antifreeze should be added for each 100 feet of pipe used. All you have to do is multiply the 7.8 gallons by four (you used 400 feet of pipe) to arrive at the proper volume of antifreeze to be pumped into the system. In doing the math, you find that 31.2 gallons of antifreeze will be needed for the loop, but you must add a little extra for the heat pump. Your chart should recommend the proper quantity for this purpose. It is likely to be in the neighborhood of five gallons.

Now that you know how much antifreeze to install, all that is left to do is install it. To do this, use the flushing pump and watertight container that was used during the flushing cycle. The container will be emptied of any remaining water and filled with the predetermined amount of antifreeze. Then the suction line of the pump will be kept below the level of the antifreeze at all times. The discharge pipe from the flushing pump will be connected to the inlet pipe from the loop.

A temporary drain line will be connected to the service valve on the outline side of the heat pump. This drain will allow water that is forced out of the loop by the induction of antifreeze to have a place to escape. Before you start the pump, close the isolation valve that was installed with the pump kit for the heat pump. You don't want the antifreeze running through the heat pump and out the drain line.

When the isolation valve is closed and the service valves are open, you can start the pump. The antifreeze will be pumped into the supply side of the loop and force water out of the temporary drain. Once all of the antifreeze has been pumped into the piping and the proper operating pressure is reached, you can close the services valve and cut off the pump.

Now you should open the isolation valve and start the circulating pump that was installed with the pump kit. With the water circulating, you can check the system's flow rate.

Checking the flow rate is a critical part of the installation and start up for a water-source heat pump. All that is required to perform this task is the opening of the isolation valves and the starting of the heat pump. As always, follow the recommendations provided with your unit from the manufacturer.

The flow rate is normally determined by observing the flowmeter. If for some reason there is little or no flow rate present, check out the circulating pump. It might be stuck or defective. If you remove the indicator plug from the center of the circulating pump, you will be able to see if the shaft is turning. Occasionally, the shaft will need a little help from the installer to get it going. This can be done by turning the shaft manually with a screwdriver. It might also pay to make sure that the pump is receiving the electricity it needs. You would not be the first installer to fume over a defective circulating pump only to find that the problem was a lack of power and not a bad pump.

When the flow rate is down and the circulating pump is working properly, there are more serious problems to be explored. If the piping in the loop has become kinked or crushed, flow through the piping can be diminished. This, of course, is a serious and expensive problem to solve. If you inspected the coil closely during installation and were careful in the backfilling process, this problem is not likely to occur.

Air can also cause the flow rate to be down. If you suspect an air lock, purge the system through the air vent. There shouldn't be any air in the system if the flushing cycle was done properly, but air could be your problem. Once you have bled the system of any unwanted air, you must recharge the piping with fresh water. This can be done by connecting a garden hose to the service valves and refilling the system. Make sure, however, that all air is forced out of the hose with water pressure before connecting the hose to the system.

Once you get the working pressure and flow to the desired level, don't be dismayed if you observe minor fluctuations in the system pressure. This is normal, as plastic pipe expands and contracts with temperature changes.

Putting the heat pump into full-time operation

Putting the heat pump into full-time operation is a job where the manufacturer's recommendations should, once again, be followed closely. Visually check all the work you have performed so far, and make sure the system is ready to start. Your particular heat pump might require steps that are different from those that are about to be described, but the following suggestions are common start-up procedures.

With the power to the heat pump turned on, position the thermostat fan position to the "On" position. You should hear the blower begin to operate. When the blower is running, you can balance the air flow at your supply ducts.

Set the thermostat to its maximum temperature, and then position the selector switch on the thermostat to the air-conditioning position. Hopefully, nothing will happen. Lower the thermostat setting until the heat pump kicks on. This will allow you to check the water flow and the compressor.

Refer to the specifications supplied with your heat pump and check such items as water temperature in the system, refrigerant pressures, temperature drops, and other recommendations made by the manufacturer. Then, turn the thermostat off. If you hear a hissing sound, the reversing valve in the unit is working properly. Leave the system shut down for a few minutes before continuing the start-up and check-out procedures.

After waiting about 10 minutes, put the system through the same check-out sequences for the heating portion of the unit. In other words, set the thermostat to the lowest temperature and put the selector lever in the heating position, nothing should happen. Raise the thermostat setting until the heat turns on, and proceed with the checklist provided by the manufacturer.

If your unit is fitted with auxiliary heat, and most are, turn the thermostat setting way up to make the back-up heat turn on. You might find that your thermostat has a setting for cutting on the back-up heat manually, most thermostats do.

Unless your unit is unusual, the steps described above will get it up and running. Set the thermostat to a position that will maintain a comfortable temperature and monitor the heat pump's operation over the course of several hours. Check periodically for leaks or other problems, and in a day or so, you can simply sit back and admire your accomplishment.

15

Equipment controls and components

When talk turns to equipment controls and components used with heat pumps, the discussion can take many turns. These items can involve everything from the heat pump's compressor to a tiny air valve. There are electrical items, coils, all types of valves, and more to cover when you take on the broad spectrum of controls and components for heat pumps.

As far as installing your own heat pump goes, you might never need to understand what a discharge muffler does or why a crankcase heater is needed. You might not care about electric resistance heaters until the day you need back-up heat and don't have any. While much of the information can be ignored during a standard installation, there might come a time during a troubleshooting phase when you must at least know these items exist.

A compressor

A compressor is a most important element of a heat pump. It is the piece of equipment that controls the flow of refrigerant through the system. The compressor does not work alone; it depends on other controls to get its job done, but the compressor is the key player in the game of circulating refrigerant.

Due to the construction of most compressors, there is little that can be done with them in the field. If a problem occurs with the unit, it will normally have to be shipped to an authorized repair facility. Other than for understanding the difference between types of compressors during the planning stage of what type of heat pump to buy, there is not much more that a homeowner needs to know.

When you are shopping for a heat pump, you will probably discover that you can choose between a model with a reciprocating compressor or a rotary compressor. Either type can be used with satisfactory results.

Reciprocating compressors can be used with a variety of refrigerants, and they are noted for their durability. Add to this their simple design and their ability to be efficient at high condensing pressures, and you have a good compressor.

Rotary compressors are not as common as reciprocating compressors, but they do offer at least two possible advantages. One of the most noticeable advantages is that rotary compressors can run with less noise than what is created with a reciprocating compressor. A deeper vacuum is also normally possible with a rotary compressor.

Reciprocating compressors are the type you are most likely to encounter, and they work well. If you have an interest in rotary compressors, do some comparison work. Read specifications from various manufacturers and assess the detailed information to determine which type will best suit your needs.

The compressor's helpers

I mentioned that a compressor does not work alone, and now I'd like to introduce you to the compressor's helpers. These helpers are controls that are used to direct the flow of refrigerant. The simplest of the group is the check valve.

A check valve (Fig. 15-1) is not a complicated device. It is simply an in-line valve that opens and closes to allow or block the flow within a pipe. In the case of heat pumps, check valves are spring-loaded. The valve is marked on the outside of its casing with an arrow. The arrow indicated the direction of flow. If a check valve is installed backwards, a system will not work.

Once the check valve is installed properly, refrigerant can only flow in one direction. If the refrigerant attempts to backflow in the pipe, the check valve will close and prevent any backflow from occurring.

Another of the compressor's helpers is a thermostatic expansion valve. This valve is more complex than a check valve. The thermostatic expansion valve controls the flow of refrigerant based on temperature and pressure. There are three primary components that make these valves work. For the valve to function properly, it must monitor pressure created by a remote bulb and power assembly, the pressure in the heat pump coil, and the equivalent pressure of a superheat spring.

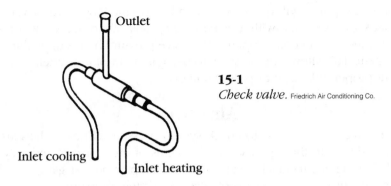

15-1
Check valve. Friedrich Air Conditioning Co.

The superheat setting of a thermostatic expansion valve is set at the factory where the valve is built. There is normally nothing that you, the homeowner, will have to do with a thermostatic expansion valve.

Another type of refrigeration control can be made from small tubing. These tubes are frequently called *capillary tubes*. While working very differently than a thermostatic expansion valve, capillary tubes perform the same basic function. They regulate the flow of refrigerant based pressure ratings.

Capillary tubes are inexpensive, and they work well when all aspects of the system remain constant. If, however, the system becomes unstable with pressure changes, problems can arise. Too much or too little refrigerant can pass through the tubes that are always open and cause the system to fail. There are pros and cons to capillary tubes. While they are the least expensive method of controlling refrigerant, they are not necessarily the best way to get the job done.

Accessory valves

There are a number of accessory valves that can be found in a heat-pump system. These valves perform a variety of duties. For example, compressor service valves allow gauges to be attached to the system for tests without shutting down the system. Air valves allow the removal of air and the checking of pressures. Relief valves are safety devices, and there are other types of valves that can play important roles in the successful operation of a heat-pump system. Let's take a look at each of these types of valves on a one by one.

Relief valves

Relief valves are safety devices that are typically required by local code enforcement offices. Paperwork supplied by the manufacturer of a heat pump will often recommend a specific type of relief valve that should be used with a given type of heat pump. Basically, a relief valve is intended to open if excessive pressure builds up within a system. This allows the pressure to be vented out of the system without property damage or personal injury.

Air valves

Air valves can be used to check system pressures. These valves are very similar to the type used in the stems of tires for cars and bicycles. A service technician can install an adapter on test gauges and use the air valve as a means of access to system pressures.

Service valves

Service valves can be used to test system pressures in place of air valves. The service valves are more expensive and take up more room, but there is less likelihood of a leak developing when service valves are used. These valves have a special port in the side of their body that allows a gauge to be screwed into it. By doing this, a technician can adjust the valve stem and take readings of the system pressure without shutting down the system. Once the testing is complete, the stem can be repositioned to close off the gauge port.

Regulating water

Regulating the flow of water with water-source systems is done to control pressure and conserve water. This feat is accomplished with the use of water-regulating valves. There is more information on water-regulating valves in chapter 14.

The magic valve

The magic valve in a heat-pump system is the reversing valve (Fig. 15-2). This valve is sometimes called a four-way valve, because it has four connection ports. What makes the reversing valve magic? This valve is used to control the direction of flow within the heat-pump system when alternating between heating and cooling requirements. Because it can sense its duty for either heating or cooling, it might appear to be magical.

Reversing valves maintain a regular flow through two of their four ports. The hot gas from the compressor accounts for one of the ports, and the suction line back to the compressor takes up the other. That leaves two ports to be dealt with. These two ports are connected to the outdoor coil and the indoor coil.

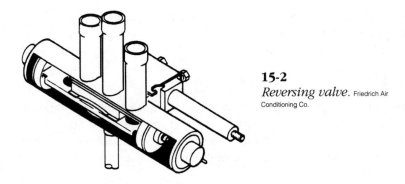

15-2
Reversing valve. Friedrich Air
Conditioning Co.

The direction of flow for refrigerant is controlled by the reversing valve based on the mode that the heat pump is operating in. In the cooling mode, refrigerant moves one way. When in the heating mode, refrigerant flows in the opposite direction. This action is made possible with the use of a solenoid-controlled reversing valve.

Other refrigerant controls

Other refrigerant controls can also be found in heat-pump systems. These controls perform various functions. For example, an accumulator is used to contain and control liquid refrigerant and oil that might otherwise make their way into the compressor. Accumulators are typically installed in the suction line at a point between the compressor and the reversing valve.

Accumulators are needed most when a heat pump is in its heating cycle. If the outdoor coil is unable to evaporate refrigerant as it should, interior flooding would be possible without the installation of an accumulator. If an air-source heat pump was in a defrosting cycle, a similar problem could arise. Due to the design of the device, an accumulator will catch excessive liquids and prevent serious damage to equipment.

A line drier

A line drier is often installed between the indoor and outdoor units of a heat pump. These driers, or dehydrators as they are also called, are mounted in the refrigerant liquid line. Any refrigerant passing through the pipe must also pass through the drier.

Dehydrators contain drying agents of one sort or another that remove moisture from the system. Many types of driers double as filters, and a number of them are made with the intent of being discarded after a set period of use. Standard procedure calls for an existing drier element to be replaced anytime the refrigerant system is opened.

Mufflers

Noise can be a problem with some heat-pump installations. When it is an installer's intent to minimize noise, a muffler is installed to reduce the discharge noise from a compressor. This concept is similar to the one used to quiet cars by muffling the exhaust.

Crankcase heaters

Crankcase heaters are often used to stop refrigerant from being absorbed into the oil contained in the crankcase of a compressor. When a heat pump is not running, refrigerant will make its way into the crankcase. During cold seasons, this can make it very difficult, if not damaging, for the heat pump to start. By keeping the crankcase warm, the refrigerant doesn't mingle with the oil in a way that will lead to problems.

Some heat pumps don't use crankcase heaters. Instead, their crankcase area is kept warm by a trickle charge of power running through windings in the compressor's motor. If you don't know which method is employed by your heat pump, check the specifications that came with your unit.

Coils

Both water-source heat pumps and air-source heat pumps depend on coils to heat and cool homes. There is one coil in the inside unit and another coil in the outside unit. The coils are sometimes called *heat exchangers,* and they are usually made of metallic pipe that is covered with aluminum fins. The tubing that makes up the coil could be made from aluminum or copper. The aluminum fins on the coils are easily damaged, and caution should be used to protect them at all times. Damaged fins result in a less effective heat pump.

Electrical controls

Electrical controls are responsible for the starting, running, and stopping of heat pumps. Without electrical controls, not much would happen with a heat pump.

Thermostats

Thermostats (Figs. 15-3 through 15-6), as you probably already know, are the control that tells a heat pump when to turn on and when to turn off. The options available in thermostats are vast. It is also possible for a heat pump to be controlled by more than one thermostat. For example, you might have one thermostat in your home, and another thermostat in the outdoor unit that controls the back-up heat. In the case of an air-source heat pump, another thermostat might be used to control the defrosting cycles.

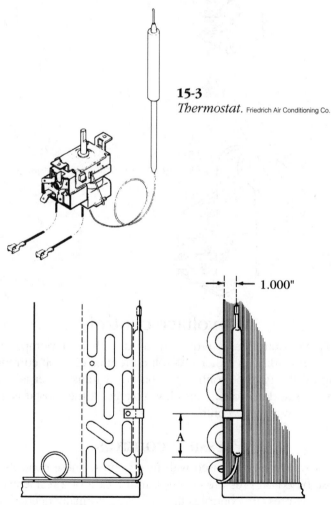

15-3
Thermostat. Friedrich Air Conditioning Co.

15-4 *Defrost thermostat bulb location.* Friedrich Air Conditioning Co.

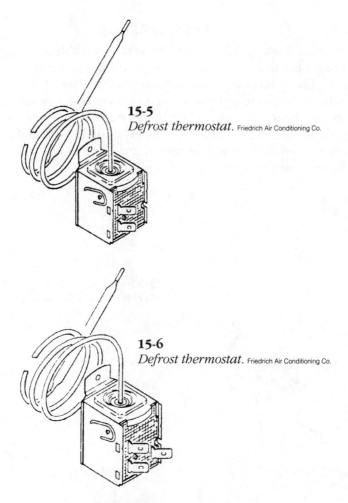

15-5
Defrost thermostat. Friedrich Air Conditioning Co.

15-6
Defrost thermostat. Friedrich Air Conditioning Co.

Low-voltage controls

Low-voltage controls are frequently used with heat pumps. These controls can work with either 110-volt current or 220-volt current, depending on the type of control purchased. A transformer is used to convert the standard voltage into low voltage. A thermostat is an example of a low-voltage control.

Pressure controls

Pressure controls are also used with heat pumps. These controls open and close based on pressure changes. In the case of a residential heat pump, a low-pressure control is the only type normally needed. However, high-pressure controls also exist and might be used on some types of heat pumps.

Motors

The motors used on heat pumps are not normally homeowner-friendly. In other words, there is not much an average person can do to work with the motor of a heat pump. Air-source heat pumps commonly have three motors involved in their operation. One of the motors is responsible for running the compressor, one is required to circulate air over the outside coil, and a third is used to run the blower on the inside unit. A water-source unit will likely have only two motors, one for the compressor and one for the inside blower. All of these motors are induction motors. If you experience problems with the motors in your heat pump, call a qualified professional to service your equipment.

Matching controls to your equipment

Matching controls to your equipment is not a job to be taken lightly. Using an incompatible thermostat can give you hours of trouble before you discover what the real cause of the problem is. Not all heat pumps are the same, and not all controls will work on all heat pumps. Matching controls and heat pumps is best done by following the recommendations made by manufacturers.

If you are buying all of your equipment from a single, reputable supplier, you should not run into problems with mismatched parts. However, if you are trying to salvage existing materials, such as a thermostat, from an old system you are replacing, problems could arise. Buying used equipment or purchasing one piece of equipment from one supplier and other piece of equipment from another supplier can put you up against some difficult situations.

Before you begin mating controls to your major equipment, make sure they are the right controls. For example, one relief valve looks pretty much like the next relief valve. Their physical appearances can be very deceptive. The relief valve for your heat pump will be required to have a specific pressure rating. A relief valve with a rating below the recommended rating will blow off before a safety interruption is required by your system. On the other hand, a relief valve with a rating higher than the specified rating for your system won't release built-up pressure until the pressure has exceeded the safe-operating pressure for your system.

In the case of relief valves, there is a little tag that dictates the pressure rating. Reading the tag will tell you if the relief valve is of the proper rating for your system. Not all parts are so easy to identify. Spend enough time reading the recommendations and installation instructions that will come with your heat pump to know that you are installing suitable controls.

Sensitive objects

Some controls are very sensitive objects. A marginal mistake in a setting on a control can render a heat pump useless, or at the least, inefficient. When you are working with controls, make sure you understand them and their operation well.

I mentioned earlier how a check valve installed backwards would not allow a normal flow. It might seem unlikely that you could make such a mistake, but I've seen numerous jobs where professional installers have installed the check valves backwards. Even though there is a direction-of-flow arrow on the outside of the valve, installing the valves backwards is not as uncommon as you might think.

The following is an array of illustrations that might help you to identify various parts of your heat pump (Figs. 15-7 through 15-17). If you are having trouble with a control, you can check in chapter 17 for advice on what the problem might be.

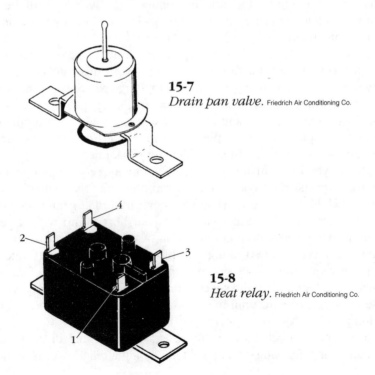

15-7
Drain pan valve. Friedrich Air Conditioning Co.

15-8
Heat relay. Friedrich Air Conditioning Co.

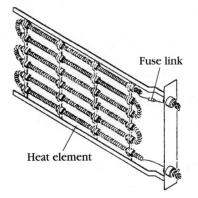

Fuse link

15-9
Heating element. Friedrich Air Conditioning Co.

Heat element

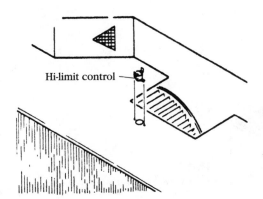

Hi-limit control

15-10
High limit control. Friedrich Air Conditioning Co.

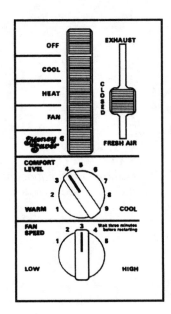

15-11
Five-button control panel.
Friedrich Air Conditioning Co

Oil ports

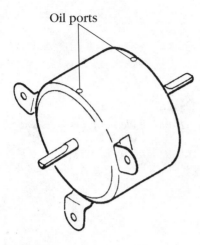

15-12
Fan motor. Friedrich Air Conditioning Co.

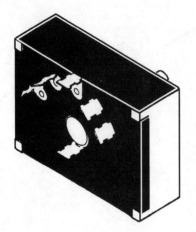

15-13
Fan speed switch. Friedrich Air Conditioning Co.

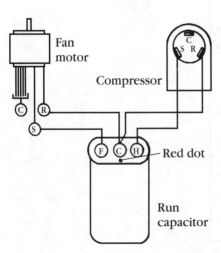

Fan
motor

Compressor

15-14
*Dual rated run capacitor
hook-up.* Friedrich Air Conditioning Co.

Red dot

Run
capacitor

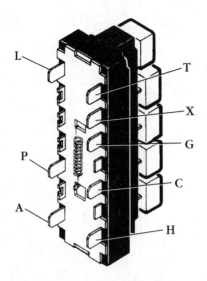

15-15 *Five-button system control switch.* Friedrich Air Conditioning Co

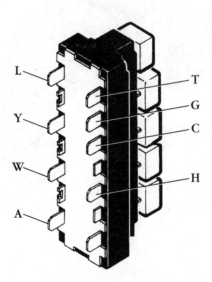

15-16 *Five-button system control switch.* Friedrich Air Conditioning Co.

" EL" series chassis parts

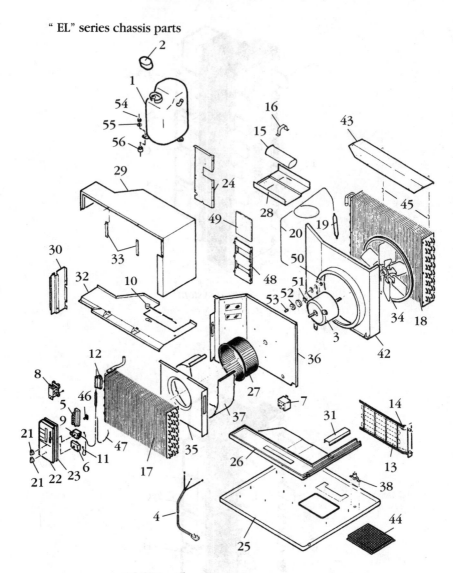

15-17A *Chassis parts.* Friedrich Air Conditioning Co.

"EL" series cabinet & mounting parts

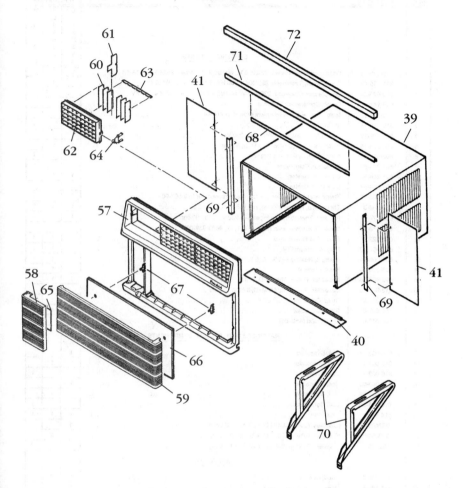

15-17B *Cabinet and mounting parts.* Friedrich Air Conditioning Co.

"EL" SERIES PARTS LIST

REF.	PART NO.	DESCRIPTION	EL19H35B	EL24H35	EL31H35B
		ELECTRICAL PARTS			
1	615-625-13	Compressor, Matsushita, 230/208 V., 60 Hz., 1 Ph., Model 2K24S3R236A-6B	1		
1	611-935-12	Compressor, Tecumseh, 230/208 V., 60 Hz., 1 Ph., Model AW5524F		1	
1	609-825-29	Compressor, Copeland, 230/208 V., 60 Hz., 1 Ph., Model CRH3-0275PFV			1
2	614-147-01	Cover, Compressor Terminal	1		
3	610-714-13	Motor, Fan — for replacement order 610-714-44	1	1	
3	610-714-14	Motor, Fan			1
4	605-000-16	Cord, Electric Supply — 30 Amp., 250 Volt	1	1	1
5	606-073-15	Switch, System (Off-On) — 5 Push Button	1	1	1
6	606-072-01	Switch, Fan Speed — 5 Position	1	1	1
7	601-264-57	Contactor, Compressor			1
8	601-264-56	Contactor, Heater	1	1	
8	601-264-57	Contactor, Heater			1
9	613-503-08	Thermostat (Cool & Heat) — for replacement order 613-503-06	1	1	1
10	605-668-00	Thermostat, Heater (Klixon), Hi-Limit	1	1	1
11	606-074-01	Resistor, Thermostat Anticipator — 230 Volt, 33,000 ohms	1	1	1
12	607-113-00	Block, Thermostat Bulb (w/alum. foil plug #603-959-00)	1	1	1
12	606-122-01	Block, Thermostat Bulb	1	1	1
*	603-959-00	Plug, Aluminum Foil	1	1	1
13	605-069-01	Element, Heating — 230 V., 5.2 KW	1	1	1
14	606-200-00	Fuse, Thermal	1	1	1
15	613-803-06	Capacitor, Run — 35/5 MFD, 370 V.	1	1	
15	613-803-16	Capacitor, Run — 40/7.5 MFD, 370 V.			1
16	610-248-00	Strap, Capacitor	1	1	1
16	600-627-01	Strap, Capacitor	1	1	1
*	606-040-05	Thermostat (Anti-Ice)			1
		REFRIGERATION SYSTEM COMPONENTS			
17	616-002-47	Coil, Evaporator	1		
17	616-002-10	Coil, Evaporator		1	
17	616-002-42	Coil, Evaporator			1
18	616-003-05	Coil, Condenser	1		
18	616-003-06	Coil, Condenser		1	
18	616-003-25	Coil, Condenser			1
19	614-813-01	Filter-Drier	1	1	1
20	03760543	†Capillary Tube — .049 I.D. x 37-¾"–40" Long	3		
20	01389985	†Capillary Tube — .064 I.D. x 27-⅛"–30" Long		2	
20	03760513	†Capillary Tube — .049 I.D. x 28-¾"–30" Long			4
		CHASSIS PARTS			
21	614-939-01	Knob, Control	2	2	2
22	616-185-01	Panel, Control (Decorative)	1	1	1
23	614-962-01	Bracket, Control Mounting	1	1	1
24	610-278-11	Bracket, Control Compartment — Back	1	1	1

* Not Shown   † Capillary tube length will vary; flow rate is the same.

15-17C *Parts list.* Friedrich Air Conditioning Co.

"EL" SERIES PARTS LIST

REF.	PART NO.	DESCRIPTION	EL19H35B	EL24H35	EL31H35B
		CHASSIS PARTS (Cont.)			
25	610-261-29	Base Pan Assembly	1		
25	610-261-31	Base Pan Assembly		1	
25	610-261-47	Base Pan Assembly			1
26	604-546-01	Pan, Drain	1	1	1
27	600-570-02	Wheel, Blower — 9" x 4-½"	1	1	
27	600-570-01	Wheel, Blower — 10-¾" x 4"			1
28	610-644-01	Brace, Condenser	2	2	2
29	610-129-16	Cover, Bulkhead	1	1	1
30	613-223-00	Panel, Left Side (Assembly)	1	1	1
31	604-993-03	Cover, Drain Trough	1	1	1
32	610-146-09	Deck Assembly	1	1	1
33	610-209-03	Support, Bulkhead Cover	2	2	2
34	605-420-01	Fan Blade, Condenser	1	1	1
35	610-224-13	Front, Blower	1	1	1
36	615-948-00	Inner Wall Assembly	1	1	1
37	610-249-01	Wrapper, Blower	1	1	1
38	601-799-00	Valve, Drain Pan	1	1	1
39	613-975-05	Shell Assembly (Cartoned with Sill Plate & Logo)	1	1	1
40	611-098-03	Sill Plate	1	1	1
*	601-691-03	Script for Shell (Friedrich)	1	1	1
*	910-029-00	Speednut for Script	2	2	2
41	602-944-05	Wingboard	1	1	1
42	606-620-00	Shroud, Condenser (Plastic)	1	1	1
43	606-619-00	Cover, Condenser Shroud	1	1	1
44	610-232-01	Screen, Base Pan			1
45	604-201-01	Clip, Condenser Shroud Cover	3	3	
45	604-201-02	Clip, Condenser Shroud Cover			3
*	603-018-00	Clip, Wire Fastening (Plastic)	2	2	2
46	614-941-01	Slide — Fresh Air & Exhaust Door Control	1	1	1
47	615-906-02	Cable — Fresh Air & Exhaust Door Control	1	1	1
48	615-840-00	Slide — Fresh Air & Exhaust Door	1	1	1
49	615-841-00	Door — Fresh Air & Exhaust	1	1	1
50	01237650	Nut — Fan Motor Mount	3	3	3
51	606-406-00	Retainer Cup — Fan Motor Mount	3	3	3
52	606-405-00	Grommet (Biscuit) — Fan Motor Mount	3	3	3
53	01336910	Sleeve — Fan Motor Mount	3	3	3
54	910-023-00	Nut (Acorn) — Compressor Mount	3	3	3
55	911-005-00	Washer (Flat) Steel — Compressor Mount	3	3	3
56	610-289-00	Grommet — Compressor Mount	3	3	
56	01150934	Grommet — Compressor Mount			3
*	614-942-05	Front, Decorative (Complete)	1	1	1
57	614-927-01	Frame, Front Decorative	1	1	1
58	614-933-01	Door, Control	1	1	1

* Not Shown

15-17D *Parts list.* Friedrich Air Conditioning Co.

"ES" SERIES PARTS LIST

REF.	PART NO.	DESCRIPTION	ES1 1H3 3A	ES1 3H3 3A	ES1 5H3 3A
		CHASSIS PARTS (Cont.)			
28	606-106-03	Wheel, Blower — 7-3/32" x 3-9/16"	1		
28	606-106-01	Wheel, Blower — 8-3/8" x 3-5/8"		1	1
29	615-740-00	Top Cover Assembly	1	1	1
30	615-733-00	Panel, Right Side (Assembly)	1	1	1
31	615-735-02	Deck Assembly	1	1	1
32	605-420-02	Fan Blade, Condenser	1	1	1
33	610-226-09	Front, Blower	1	1	1
34	615-738-00	Inner Wall Assembly	1	1	1
35	615-225-01	Wrapper, Blower	1	1	1
36	601-799-00	Valve, Drain Pan	1	1	1
37	613-975-08	Shell Assembly (Cartoned with Sill Plate & Logo)	1	1	1
38	611-098-01	Sill Plate	1	1	1
*	601-691-03	Script for Shell (Friedrich)	1	1	1
*	910-029-00	Speednut for Script	2	2	2
39	602-944-06	Wingboard	1	1	1
40	606-620-03	Shroud, Condenser (Plastic)	1	1	1
*	603-018-00	Clip, Wire Fastening (Plastic)	2	2	2
41	614-941-01	Slide — Fresh Air & Exhaust Door Control	1	1	1
42	615-906-00	Cable — Fresh Air & Exhaust Door Control	1	1	1
43	615-841-00	Door — Fresh Air & Exhaust	1	1	1
44	615-840-00	Slide — Fresh Air & Exhaust Door	1	1	1
45	01237650	Nut — Fan Motor Mount	3	3	3
46	606-406-00	Retainer Cup — Fan Motor Mount	3	3	3
47	606-405-00	Grommet (Biscuit) — Fan Motor Mount	3	3	3
48	01336910	Sleeve — Fan Motor Mount	3	3	3
49	610-289-00	Grommet — Compressor Mount	3	3	3
50	914-004-00	Screw — Compressor Mount	3	3	3
*	614-942-03	Front, Decorative (Complete)	1	1	1
51	614-925-01	Frame, Front Decorative	1	1	1
52	614-931-01	Door, Control	1	1	1
53	614-928-01	Center Section (Intake Grille)	1	1	1
54	614-936-00	Louver, Vertical	18	18	18
55	614-936-01	Louver, Vertical Control	3	3	3
56	614-934-01	Grille, Discharge Air (Horizontal Louvers)	3	3	3
57	610-787-02	Connector, Louver	3	3	3
58	614-938-00	Pivot Piece, Discharge Grille	2	2	2
59	614-943-01	Label, Operating Instructions	1	1	1
60	608-658-06	Filter, Air (9-1/2" x 16-1/2" x 3/8")	1	1	1
61	608-659-00	Holder, Filter	2	2	2
*	611-050-00	Accessory Package	1	1	1
62	611-097-01	Angle, Wingboard (Top)	1	1	1
63	611-096-01	Angle, Wingboard (Side)	2	2	2
64	611-095-03	Bracket, Support	2	2	2

* Not Shown

15-17E *Parts list continued.* Friedrich Air Conditioning Co.

"ES" SERIES PARTS LIST

REF.	PART NO.	DESCRIPTION	APPLICATION		
			ES11H33A	ES13H33A	ES15H33A
		CHASSIS PARTS (Cont.)			
65	606-103-02	Gasket, Wingboard Seal (Vinyl)	1	1	1
•	608-460-05	Hardware Assembly (nuts, bolts, etc.)	1	1	1
	01311760	Screw — #8A x ⅝" Hex. Head (for Plastic Front)	2	2	2
66	600-733-00	Gasket, Window (Foam)	1	1	1
•	616-306-02	Carton, Sleeve	1	1	1
•	616-307-01	Liner, Carton	1	1	1
•	616-308-01	Pad, Shipping (Top)	1	1	1
•	616-321-02	Pad, Shipping (Front)	1	1	1
•	616-246-02	Pad, Shipping (Bottom)	1	1	1
		OPTIONAL ACCESSORIES			
•	01900-235	Drain — Condensate Connection Kit, DC-2	x	x	x
•	610-089-03	Start Kit, Capacitor/Relay (Pow-R-Pak)	x	x	x
•	616-238-01	Winter Cover	x	x	x

• Not Shown

15-17F *Parts list continued.* Friedrich Air Conditioning Co.

16

Upgrading existing heat-pump systems

Deciding on when upgrading an existing heat-pump system makes sense can be confusing. It stands to reason, however, that there comes a time when an existing heat pump is not as cost-effective to operate as a newer model would be. Technology in this world changes quickly, outdating what just a few years ago was a top-of-the-line model of equipment. Whether you are talking about computers, cameras, or heat pumps, newer models keep coming out that are better than the equipment they replace. The question is, when does it make sense to replace your old heat pump with a new one? A similar question could be posed in regards to any type of existing heating or cooling system. If you own a home that is equipped with heating and cooling equipment that has a good bit of age on it, you might find it worthwhile to upgrade your system.

Cost considerations

Cost considerations are almost always a factor in deciding to upgrade an existing heating or cooling system. When you think about the cost of doing the work, you must weigh more than just the acquisition cost of the new equipment. To begin with, you have to decide how much you can save in operating expenses and how long it will take to recover your investment. If you are gaining the benefit of air conditioning that your present forced hot-air furnace doesn't give you, an adjustment in the initial cost must be factored in to allow for the added benefits gained from a heat pump that will both heat and cool your home. Before you are done looking at all the numbers, you might develop red eyes and a headache. To help simplify the mystery of cost considerations, here is a list of them, one by one.

Initial cost of conversion

The initial cost of conversion is usually the first cost factor a homeowner addresses when thinking of installing a new heat pump. This is a reasonable place to start, because the out-of-pocket expense might be so much that the job never materializes.

You've already learned what to look for in a heat pump and how to look for it. Earlier chapters helped you to understand the differences between various heat pumps and how these differences can affect the cost of a unit. Now it is time to put that knowledge to work.

Using the information gained from this book, you should be able to plan the installation of a new heat pump pretty well. You should also be able to project potential problems that might increase the apparent acquisition and installation costs. An example of such a circumstance could be that the duct work for your old forced hot-air furnace is too small to use with a modern heat pump. When you are putting together prices for a new heat pump, it is important to include a figure for all the costs you are likely to incur. It also makes sense to add a little to your estimate to cover expenses that might have been overlooked.

Once you have pulled together a complete package of materials needed to do the job, you must assess your ability to perform the installation. Again, this book should have given you enough information on what is involved with an installation and conversion to make it easy for you to decide on how much of the work you can do yourself. Chapter 19 expands on this issue and helps you determine what your limitations are when it comes to installing a heat pump.

When you have a clear picture of what the new equipment will cost and how much professional help will be needed, you are ready to move into the next phase of the evaluation process. This is where most people have trouble sifting through the pros and cons.

Recovery period

The recovery period for recouping your initial investment is of paramount importance in deciding to upgrade an existing heating or cooling system. It doesn't make much sense to lay out thousands of dollars to get a more efficient heat pump if it will take 10 years to recover your investment and you know you will be moving in three years. The new equipment might add some value to your house when you sell it, but then again, it might not add nearly enough to re-cover your costs. For this reason, it is wise to operate on known facts.

There is no doubt that a heat pump is one of the most efficient means with which to heat and cool a home. Heat pumps, when de-signed and installed properly, can save a homeowner a lot of money in a relatively short period of time. However, not all conversion pro-jects are the same, and each one must be looked at on an individual basis. To do this, you must know a little about efficiency ratings.

Efficiency ratings

Efficiency ratings can be used in making a decision on whether or not it will be cost effective to install a new heat pump to replace an ex-isting system. When we talk about heat pumps, the efficiency rating is known as the coefficient of performance (COP). In the case of an oil-fired heating system, the efficiency rating is simply that, an effi-ciency rating. Here's an example that will help you to see how effi-ciency ratings can make it easier for you to commit to upgrading to a new heat pump.

Assume that you presently have an oil-fired, forced hot-air fur-nace. You would like to have a new heat pump, because you want air conditioning and you've heard that heat pumps can save you money. There is no question that the heat pump will provide you with air conditioning, and this is a benefit you must weigh separately from the basic cost of converting your heating system.

The first thing that you should know is that an oil-fired furnace never runs at 100% efficiency, even when its brand new. Most oil-fired furnaces and boilers struggle to operate at efficiency rates in the low eighties. In other words, if you were to install a new oil-fired heating system, the unit would probably be only 80%, give or take, efficient. This means you are wasting a significant amount of oil in lost efficiency. Old oil-fired units might have efficiency ratings of 73% or less. Now, how does this fit into the question of buying a new heat pump.

Heat pumps are rated in terms of COP. Let's assume your existing oil-fired heating system is of average condition, but only 60% effi-cient, and that the price of oil is one dollar a gallon. Further assume that you use 1,000 gallons of fuel oil each year to heat your home. The total cost of your heating oil is $1,000. There is also some cost for electricity that is used to run the blower or circulator, but to keep this example simple, we won't factor in that cost. You know the efficiency rating of your oil-fired unit is poor, so you are considering replacing it with a new heat pump.

Now we have to assign a cost to electricity for running the heat pump that will replace the oil-fired unit. Keep in mind that oil prices and electric prices fluctuate and that the figures used in this example are for illustrative purposes only. The heat pump you are considering has a COP rating of three. This means it will produce three times the amount of heat that electric heat would with the same amount of electricity used.

One kilowatt hour of electricity is equal to 3,413 BTUs of heat. Your current oil usage translates into 24,000 kilowatt hours of electricity. Assume that the cost of one kilowatt hour of electricity costs eight cents. This would mean that 24,000 kilowatt hours would cost $1,920. How much is it going to cost to operate the new heat pump? Because the heat pump has a COP of three, you divide the $1,920 by three to obtain the annual cost of electricity to run the heat pump. The correct answer is $640.

Your oil-fired monster costs $1,000 a year in oil, plus the cost of electricity. Working strictly off the cost of oil, the heat pump will cost $360 a year less to operate. When you factor in the electrical usage of the oil-fired unit, your savings are even greater. Okay, so let's say that your overall annual savings by converting to the heat pump will be about $475. Let's further assume that the acquisition and installation cost of a new heat pump will be $4,750, I've picked this number only to make the math easy to understand. It would take 10 years for your new heat pump to pay for itself, based on the fuel prices used in the equation. Assuming that the spread between fuel prices remain static, you will begin to save money after the 10th year. If you are fanatical on details, you might want to compute how much interest you would earn on the $4,750 during the first ten years and extend the payback period to cover the loss of your interest earnings. At any rate, I think this example makes it easy to see how to use efficiency ratings to gain more insight into your conversion plans.

Suppose your house was equipped with electric baseboard heat, how would you evaluate the installation of a new heat pump? If the heat pump you are considering has a COP of three, it will take one-third less electricity to run it than it will to run the electric baseboard heat. If your annual heating cost for the baseboard heat is $1,200, the cost of running the heat pump would be $400. We are comparing heating cost only. If you run the air conditioning from the heat pump during the summer, obviously the total operating costs will be more than what has been shown in these examples.

Heat pump to heat pump

If you are switching from one heat pump to another heat pump, you need only be concerned with the COP ratings of the two units and the overall condition of your existing unit. This is an apples-for-apples evaluation that is easy to do. If your old heat pump has a COP of two and the new heat pump has a COP of three, you can project the payback period easily. To illustrate this, let's look at a quick example.

Assume your annual kilowatt usage with the existing heat pump is 15,000 hours at eight cents per hour. This translates into an annual expense of $1,200. The new heat pump would do the same job for $800. You would save $400 a year by replacing the old heat pump.

To arrive at this answer, I assumed an annual kilowatt usage of 30,000 hours. Because the old heat pump has a COP of two, I divided the 30,000 hours in half and came up with 15,000 hours. 15,000 hours multiplied by eight cents per hour gave me $1,200. When I studied the new heat pump, with a COP of three, I divided the 30,000 kilowatt hours by three and came up with 10,000 hours. 10,000 hours multiplied by eight cents an hour gave me $800 and a difference of $400.

Money is not everything

Money is not everything, but it does have a lot of meaning in our society. Very few homeowners would go to the trouble and expense of installing a new heat pump just for the fun of doing the job. Most people either want to save money on their heating and cooling expenses, are forced to replace a failing system, or they want the added benefit of air conditioning that they don't presently have. All three of these reasons are good cause for considering a new heat pump.

Air conditioning

Air conditioning is a very nice amenity to have in most parts of the country. Even in Maine, air conditioning comes in handy for a few weeks out of each year, and in many states, life without air conditioning is pretty miserable. It is hard to put a price on the value of central air conditioning. Oh, real estate appraisers don't have any trouble assigning a value to the benefit of air conditioning, but when cool, comfortable, conditioned air is measured on a personal level, its value varies considerably.

I hate hot, humid weather conditions. For me, air conditioning has a tremendous value. There are a number of people who don't share my love or need for air conditioning. When you are evaluating the cost-effectiveness of a new heat pump, you will have to make a personal decision on how much the air-conditioning feature is worth to you. It might be worth the total cost of the conversion work, or it might not be worth much at all. In any case, the personal value placed on air conditioning cannot be ignored when evaluating the feasibility of installing a new heat pump to replace some type of older heating system.

Disruption

The disruption caused in a home that is undergoing a conversion project can be substantial, especially if the home is not equipped with ducts that are suitable for the new heat pump being installed. People who are allergic to dust might have to move out of their home while much of the work is being done. Not only is a situation like this annoying, it can get expensive. You should consider the chaos that your home could be in during the conversion project. In most cases the disruption will be justified, but there are some people who would rather live with an outdated heating system than endure the commotion caused by converting to a heat pump.

Your personal reasons for wanting to install a new heat pump can outweigh the costs. Even if you find that the job is not cost-effective, you might have reasons that will justify the job. Don't overlook your personal feelings when considering the cost of the project.

A methodical approach

A methodical approach is the best way to decide if you should invest in a new heat pump. Look at the hard costs. Factor in any additional benefits you will derive from a new heat pump, such as air conditioning. List your personal reasons for wanting to invest in new equipment. Spend enough time on each of these aspects of the decision-making process to be fair to yourself. If you move from category to category in an organized manner, it will be easier to draw a clear conclusion to your consideration of a new heat pump. There are a few more factors to put under the microscope as you go along in your decision-making process.

Can you sell your old equipment?

Can you sell your old equipment and use the money to offset the cost of your new heat pump? If your present system is in good working order, you probably can. There is always a market for good, used equipment. You might even find a supplier that will accept your old equipment as a trade-in on the new heat pump; it's worth asking about.

Can you build equity in your home?

Can you build equity in your home by improving the heating and cooling system? You might be able to. The answer will lie in your ability to do most of the work yourself and the local real estate market. If your home is one of the few in your neighborhood to not have a heat pump, installing one could very well be worthwhile. However, you might find that the new system does very little to increase the value of your home, especially if heat pumps are not common in your area. You might even hurt the resale value of your home by installing a heat pump. Some areas, like Maine, believe forced hot-water heat is the best type of heat to have in a cold region. Because most houses in Maine have hot-water baseboard heat, trying to sell a house with a heat pump could be difficult. If you want to know for sure how a heat pump will affect the value of your home, you should consult a licensed real estate appraiser.

How difficult will the conversion be?

How difficult will the conversion from your present heating system to a new heat pump be? If your home is not equipped with ductwork, the job might be more than you are willing to deal with. Earlier chapters explained how to make these types of evaluations, but you should not overlook the burden of the conversion when you are assessing the feasibility of the job.

The bottom line

The bottom line when it comes to a heat-pump conversion rests on your shoulders. It is your house, your money, and your time. Only you can decide when the time is right to upgrade your existing heat pump or convert to a new heat-pump system.

17

Troubleshooting heat pumps and their components

This chapter is all about troubleshooting heat pumps and their components. If you've ever had to endure a sweltering summer day or a cold winter night when your heat pump wouldn't work, you'll appreciate the value of this chapter. For those of you who have been lucky enough not to experience a heating or cooling failure, there is no time like the present to prepare for the unexpected.

Many people don't associate troubleshooting with the installation of a new heat pump. In their minds, troubleshooting and repair work is something that is done when heat pumps age and become less dependable. The truth is, a lot of troubleshooting and repair work is done in conjunction with new installations.

Any professional installer will be able to tell you a long list of stories about new parts that were defective and new equipment that simply didn't do what it was supposed to do. Anyone who has worked in the field for very long will have had their share of problems with new installations. Because the heat pump you plan to install will probably be the first and last heat pump you ever install, your need for easy-to-understand, effective troubleshooting techniques is of paramount importance.

Calling a service technician at any time is an expensive proposition, but calling one at night or on a weekend, when most homeowners do their around-the-house work, gets extremely expensive. With a little skill, a lot of patience, and the help of this chapter, you might be able to avoid the high cost of a professional service technician.

As I'm sure you know by now, you should always consult and adhere to recommendations made by the manufacturer of your heat pump. Most, if not all, heat pumps are supplied with instructions for troubleshooting problems that might occur with the unit. Those suggestions, coupled with this chapter, will more than prepare you for finding the causes of your problems.

Troubleshooting techniques for air-source heat pumps are somewhat different from the procedures used with water-source heat pumps.

Troubleshooting air-source heat pumps

Troubleshooting air-source heat pumps is not difficult if you have enough data to consult. In many ways, troubleshooting is a trial-and-error procedure. Professionals learn how to minimize wasted time by avoiding unneeded troubleshooting steps. While this is fine for professionals, it is often best for inexperienced people to take the troubleshooting process slowly. This might mean following step-by-step instructions that fail to be fruitful, but in the end, the right moves will be made. If you, as an inexperienced installer, try to skip through certain steps, you could wind up losing more time than if you had gone by the book, so to speak.

One of the worst problems you are likely to face with your new heat pump is that it simply won't run. The frustration level reached when a new unit has been installed and won't run can be extreme. What would you do first if your new heat pump wouldn't turn on? Well, kicking it won't solve your problem, so let's see what will. If you will be using a water-source heat pump, you can skip ahead to the next section that deals with them.

Won't run

If you have a unit that won't run at all, there is probably a very simple solution to the problem. In many cases, the problem is a lack of electricity. A fuse might be blown or a circuit breaker might be tripped. Perhaps you had turned off the power while you were working on the installation and failed to turn it back on. Maybe one of your children moved the disconnect lever when you weren't looking. At any rate, check all fuses, circuit breakers, and disconnect switches to make sure they are as they should be.

If everything is normal, check the voltage in the condensing unit. Don't attempt this step unless you are familiar with working with electricity and electrical meters. These troubleshooting steps will, in many cases, involve working with electricity. The power from electrical wiring can be deadly. If you are not skilled in working with electricity, defer to a licensed professional for all work involving electrical testing. If you find that the proper voltage is present for the condensing unit, you are working your way through the many possibilities for the cause of your problem.

Before you go too far with technical tests, try a very simple test. Go to your indoor thermostat and check all of the settings. Is the selector lever set on the proper cycle? In other words, if the selector is set for cooling on a cold winter day, the heat pump is not going to run. Has the thermostat been set at a temperature setting that would not call for the heat pump to turn on? For example, if the selector is set in the heating mode and the thermostat temperature setting is below the present temperature in the home, the heat pump is not going to work. These are extremely simple problems to check for, but they are just as capable of keeping the heat pump from running as some serious types of defects.

Assuming that all of the settings for the thermostat are set properly, check the control to see if it is out of calibration or defective. Your new thermostat should have been packed with its own set of instructions and troubleshooting tips. Due to the variances between thermostats, it is important that you follow the instructions given for your particular model.

If you have a high-pressure control, and you probably don't, it could be stuck in the open position. Refer to the manufacturer's paperwork and check the control.

A bad transformer can keep your heat pump from starting. If you understand electrical wiring, you should check to see that the transformer is wired correctly. Consult the manufacturer's recommendations for troubleshooting the transformer and follow them. It might be necessary to replace the transformer.

It is possible that the cause of your problem, if you haven't found it by now, is the contacts in the compressor. External overloads can be overcome by replacing the contacts. Internal overloads require that the compressor be replaced.

If you are dealing with new equipment, the cause of your problem is most likely a simple one. Take your time in checking all the possible causes, and you should have very little trouble finding the root cause of your problem.

Low liquid pressure

A reading of low liquid pressure and a unit that will not cool properly is a sign of trouble. The cause of the trouble can be one of many. There might not be an adequate amount of refrigerant in the system. If the system is low on refrigerant, it would be wise to check the entire system for leaks.

The problem could be caused by something as simple as a dirty filter or coil. Inspect the coil and filter to see that they are clean and allowing a clear air flow.

A unit that is not cooling well and that is giving a reading of low liquid pressure might be caused by check valves that are installed backwards. Inspect the check valve to see that it is installed in the right direction. If it's installed properly you might ultimately have to check to see that it is working correctly. A bad expansion valve can also cause a unit to give the same trouble symptoms caused by a bad check valve.

There are two other possible causes that are capable of causing a heat pump to cool poorly and produce a reading of low liquid pressure. A restriction in the liquid line could be at fault. If that's not the case, you should divert your attention to the compressor valves. Rely on the instructions provided by the manufacturer of your equipment in working with the compressor.

Because heat pumps heat homes as well as cool them, you must consider the problem of having a low liquid pressure and insufficient heat, as opposed to cooling. When this is the case, you have to change your troubleshooting procedure.

If the system does not have a suitable charge of refrigerant in it, you could experience a low liquid pressure and a lack of satisfactory heat. Check the refrigerant and add some if necessary. When refrigerant is low, there is a chance some part of the system has a leak in it. You should check the system out thoroughly to discover any hidden leaks.

Restrictions in the refrigeration lines can also contribute to a low liquid pressure and a poor heating experience. Look over your refrigerant lines and see if any visible restrictions are present.

The valves in the system could also be at fault. Inspect the check valve and expansion valve. Make sure the check valve is installed with the arrow on the outside of the body pointing in the direction of the intended flow.

If you are particularly unlucky, you might find that the compressor valves are bad. This will probably call for a replacement compressor, but as always, refer to the suggestions provided by the manufacturer of the equipment.

High liquid pressure

A unit that is producing a reading of high liquid pressure and failing to cool air satisfactorily might be suffering from a dirty outside coil. Because this is an easy cause to detect or rule out, it is the most sensible step to take first. Inspect the outside coil and flush it out if necessary.

If the coil is not responsible for your problem, there are some other aspects of the outside unit you should check. Observe the fan in the outdoor unit to see that it is turning in the proper direction. Also, check to see that there are no loose blades on the fan. Obviously, you must be careful not to get fingers caught in a moving fan, so take necessary precautions when working around the blade.

Once you have ruled out the coil and the fan, check out the motor. Is it running at the proper revolutions per minute (RPMs)? Is the run capacitor defective? After you have investigated the motor, move on to the refrigeration lines.

If the outside unit is not at fault, check the charge of refrigerant in the system; it might be overcharged. There is also a chance that noncondensables have found their way into the system and will have to be purged out.

Now that we have covered the troubleshooting for the cooling side of the unit, let's talk about the same problem on the heating side. When the unit has a high liquid pressure and is not heating properly, check the indoor coil and filter. If these items are dirty, they might be causing your problem.

If the filter and coil checks out okay, turn your attention to the refrigerant charge. If the system is overcharged, you might have found the cause of your problem.

Noncondensibles in the system can also cause a heat pump to heat poorly and to run with a high liquid pressure. If the system has been tainted by noncondensibles, you will have to purge them from the piping.

If the blower motor on the heat pump is not sized properly or is running backwards, you could consider it the cause of your problem. Check the motor to ensure it is of the proper size, is running at the right speed, and is turning in the correct direction.

Heat registers that are closed could be your only defect. Improper air flow can cause a high liquid pressure and a lack of heat. Check the registers to be sure that they are open and unobstructed. At the same time, check the return grill to make sure it is not blocked.

If you have connected your new heat pump to existing duct work that served some other type of equipment previously, there might be a problem with the size of the duct work. This, of course, could be a major problem. If you have ruled out all of the other possibilities, you might have to delve into the duct work and check its sizing.

Low suction pressure

What should you look for when your unit has low suction pressure and is not cooling properly? One of the first places to look is at the filter. If the filter is dirty and blocking air flow, your heat pump could experience low suction pressure and a lack of cooling ability. Any restriction in air flow could cause this type of problem.

If refrigerant is not being returned to the compressor as it should be, the heat pump will exhibit low suction pressure and poor cooling performance. Test the refrigerant charge and consider looking for leaks if the charge is low.

A defective expansion valve or a fan in the outside unit that is running backwards could be the cause of the problem. While it is unlikely, you might find that the motor for the fan is not sized properly. A defective run capacitor could also be the culprit. Any of these causes could be the root of your problem.

If you are suffering from low suction pressure and a lack of heat, check the outside coil to see if it is dirty or obstructed. While you are outside, check the fan and the motor on the outside unit. Make sure the fan is not running backwards and that the motor is running at the proper speed. Check the run capacitor to see that it is doing its job.

Other causes of low suction pressure and a lack of heat could be defective expansion valves, restricted refrigerant lines, or improper refrigerant charges.

High suction pressure

Suppose the heat pump is not cooling efficiently and is showing signs of high suction pressure, what should you do? Testing the refrigerant charge might prove that the system is overcharged. You might find that the check valve or the reversing valve is defective or installed improperly. There is also a possibility that noncondensibles have invaded the system. As something of a worst-case scenario, the compressor valves might be bad.

When you are experiencing high suction pressure and a lack of heat, you might find that the refrigerant lines are overcharged. There is also a possibility that noncondensibles are infecting the system. A bad check valve or reversing valve could also be at the root of the problem. And, it is possible that the compressor valves are bad.

Operating pressures are normal

When operating pressures are normal and a heat pump will heat or cool properly, the list of possible causes is not so long. Assume that you have a heat pump that is running at normal operating pressures, but it is not producing enough cool air. What are you going to look for? Air leaks are one cause you could look for, but you should begin by looking at the electrical wiring, assuming that you are experienced enough to work with electrical wiring and components safely. Inspect all wiring, circuits, and components to see that there are no defects present. Make sure the wires are not loose and that they are connected to their terminals properly. If the wiring checks out, look for air leaks.

Air leaks aren't likely, but it is possible a section of duct work has come loose. If the installer failed to secure supplies to the trunk line or supply boots, the supply duct might have dropped out of place. This could cause cool air to be blown into some area other than the intended cooling zone.

Don't overlook the possibility that supply registers are closed. If you have young children, the fascination of closing off all the floor registers might be more than their curious fingers can resist.

If you have proved that the wiring is fine and that air leakage is not a problem, there is only one other conclusion to draw to, and it is not a pretty one. In the event that the simple checks have turned up empty, you must consider the fact that the heat pump might not have been sized properly. This is certainly not a thought you would want to conjure, but it is a possibility.

When an undersized unit is suspected, go back over the system and size it again. This is the only way to be certain that the equipment is not of the wrong tonnage. If you sized the system yourself, it might pay to call in a professional to check over your sizing calculations and installation.

If your system is running at routine pressure and is not providing adequate heat, check the outside thermostat. The cause of your problem might be as simple as a thermostat setting that does not meet your needs.

Assuming that the thermostat is not causing your lack of heat, check the duct work to see that there is proper air flow and no leaks. Again, check the registers to be sure they are open. Inspect any dampers that have been installed in the duct work to be sure that they are not restricting air flow. If any of the ducts run through unheated space, such as a crawl space, make sure that they are insulated. Even if you insulated the ducts during installation, there is a chance the insulation has come loose and fallen off.

Hot air instead of cool air

If you have your heat pump in the air-conditioning mode and are getting hot air instead of cool air, you should suspect the reversing valve. If the reversing valve is not functioning properly, it is possible to get warm air when you want cool air. Inspect the valve relay to make sure the device is working as it should.

Problems in the wiring could also cause a heat pump in the cooling mode to produce warm air. Check the amperage of the electric resistance elements. If you get an amperage reading, move to the indoor thermostat and check the wiring for shorts. If no shorts are found, you might have a defective thermostat.

Won't defrost

When an air-source heat pump won't defrost itself, problems are sure to follow. What would cause a unit to not defrost? There are two possible causes that come to mind. The reversing valve might be responsible for the problem. If the relays of the reversing valve are closed, the problem could manifest. It is also possible that the defrost controls are defective. If the pressure pipe is obstructed, such as with a kink in the tubing, the unit might not be able to defrost. Check these two possibilities and refer to your owner's manual for recommendations from the manufacturer.

Won't stop defrosting

While some heat pumps won't defrost at all, others won't stop defrosting. One simple cause of this problem is a clogged indoor coil or filter. Because this is easy to check for, it is a logical place to begin the troubleshooting process. If the coil or filter is dirty, it will have to be cleaned or replaced.

The temperature bulb that controls the defrost cycle might be loose or uninsulated. This is certainly worth a look. Defective defrost controls might be causing the never-ending defrost cycle, and they might have to be repaired or replaced. The expansion valve could also cause a problem with the defrosting cycle, so it should be checked out. If none of these causes seem to be likely, check the refrigerant charge to be sure it is at an acceptable level.

Partial defrosting

There might be times when an air-source heat pump will only obtain partial defrosting during the defrost cycle. When this is the case, there are two relatively simple causes that might be responsible and one not-so-simple cause to look out for. Let's hear the bad news first. When a unit is not defrosting completely, the compressor valves might be defective. However, before you jump to the conclusion that the compressor is shot, check the refrigerant charge to see that it is where it should be.

If there are no abnormalities in the refrigerant charge, check the defrost controls and circuits. This will require testing for pressure and temperature. Refer to recommendations of the manufacturer of your particular piece of equipment for guidance.

Compressor turns off early

If you have a heat pump where the compressor turns off early during a defrost cycle, check the indoor coil and filter. If these items are dirty or restricted, clean them, make sure good air flow is possible and test the equipment to see if your simple repair has made a difference.

Assuming that the problem is not associated to a dirty coil or filter, check the expansion valve to see if it might be defective. Check for pressure and temperature, as per the manufacturer's recommendations.

Your problem could be caused by a bad check valve or even the reversing valve. Inspect these valves to make sure they are installed and operating properly.

Auxiliary heat

The auxiliary heat in your heat pump might stay on after a defrost cycle is completed. If this happens, the thermostat for the heat is the most logical place to begin your troubleshooting. With the heat pump in a defrost cycle, check out the thermostat. You might have to replace it.

Another cause for the auxiliary heat not turning off could be a defective relay. If you have the knowledge to work with relays, check out the one for the emergency heat. If it is defective, replacement will be required.

The components and circuitry associated with the auxiliary heat might require your attention. The three things to check are the de-ice control, the control relay, and the contactors. Field repair or replacement might be required.

Auxiliary heat won't come on

If the auxiliary heat in your heat pump won't come on during a defrost cycle, the problem is going to be caused by either the defrost components or its circuitry. To determine the exact cause of the problem, you will have to check out all of the components and their wiring. Follow the specific recommendations of the equipment manufacturer to determine which of the components is not working properly.

Troubleshooting water-source heat pumps

Troubleshooting water-source heat pumps is similar in many ways to troubleshooting air-source heat pumps. However, due to the different operating procedures for the two types of systems, there are obviously some variations in the troubleshooting process. The water-source heat pumps involves components not found in an air-source system. There are also aspects of an air-source system that are not present in a water-source system. For this reason, I cover the same basic problems for water-source heat pumps that was just studied for air-source heat pumps. There are many similarities in some circumstances, and none in others.

Won't run

If you have a unit that won't run at all, there is probably a very simple solution to the problem. In many cases, the problem is a lack of electricity. A fuse might be blown or a circuit breaker might be tripped. Perhaps you had turned off the power while you were working on the installation and failed to turn it back on. Maybe one of your children moved the disconnect lever when you weren't looking. At any rate, check all fuses, circuit breakers, and disconnect switches to make sure they are as they should be.

If everything checks out to be normal, check the voltage in for the unit. Don't attempt this step unless you are familiar with electrical meters and working with electricity. These troubleshooting steps will, in many cases, involve working with electricity. The power from electrical wiring can be deadly. If you are not skilled in working with electricity, defer to a licensed professional for all work involving electrical testing. If you find that the proper voltage is present for the condensing unit, you are working your way through the many possibilities for the cause of your problem.

Before you go too far with technical tests, try a very simple test. Go to your indoor thermostat and check all of the settings. Is the selector lever set on the proper cycle? In other words, if the selector is set for cooling on a cold winter day, the heat pump is not going to run. Has the thermostat been set at a temperature setting that would not call for the heat pump to cut on? For example, if the selector is set in the heating mode and the thermostat temperature setting is below the present temperature in the home, the heat pump is not going to work. These are extremely simple problems to check for, but they are just as capable of keeping the heat pump from running as some serious types of defects.

Assuming that all of the settings for the thermostat are set properly, check the control to see if it is out of calibration or defective. Your new thermostat should have been packed with its own set of instructions and troubleshooting tips. Due to the variances between thermostats, it is important that you follow the instructions given for your particular model.

A bad transformer can keep your heat pump from starting. If you understand electrical wiring, you should check to see that the transformer is wired correctly, has the proper voltage coming through it, and is not burned out. Consult the manufacturer's recommendations for troubleshooting the transformer and follow them. It might be necessary to replace the transformer.

If you are dealing with new equipment, the cause of your problem is most likely a simple one. Take your time in checking all the possible causes and you should have very little trouble finding the root cause of your problem.

All heat

If you have a problem where your heat pump gives you all heat and no cool air, the problem is almost guaranteed to be associated with the reversing valve. The reversing valve is probably wired incorrectly. Check the wiring at both the heat pump and the thermostat. If you are convinced there is nothing wrong with the wiring, you must troubleshoot the valve itself and might have to replace it.

Not enough cool air

If you are not getting enough cool air from your heat pump, there are several possible causes for the problem. Start with a simple check. Inspect the supply registers to see that they are open. Inquisitive children will sometimes close registers just to see if they can do it. Obviously, if the registers are closed, a sufficient volume of cool air cannot fill your home. You might also want to check the dampers in the duct work to make sure they are open and operating properly.

Another simple check can be made on the air filter. If the filter is dirty, it could be the cause of your problem. Once you get past the registers, dampers, and filter, the process gets a little more complicated.

The next thing to check is the compressor. A compressor that runs but that does not cool the evaporator coil points to either a defective compressor or a lack of refrigerant. Check the refrigerant to make sure it is at a normal operating charge. If necessary add refrigerant to the system. If the problem still exists, you either have a bad compressor or a blocked water flow in the coil. It is also possible that there is not a sufficient water flow to maintain cool temperatures.

Investigate as much of the piping as you can to see if there are any visible signs of trouble, such as a kinked pipe. If there isn't, you must establish that there is an adequate water flow. You might want to increase the flow rate, depending upon what the specifications from the manufacturer of your equipment suggests.

Won't come on

Sometimes heat pumps won't come on when there is a demand for heat. This problem is usually fairly simple to correct. To keep the troubleshooting as simple and effective as possible, check the air filter first. If it is dirty, replace it and try the unit. If it still refuses to come on when heat is called for, move your troubleshooting to the thermostat.

Before you start digging around in the wiring of the thermostat, check to see that the controls are set in the proper position. It would be silly to tear the thermostat apart only to find that it was not set in the heating mode or that it is set below the existing room temperature.

Once you have determined that the thermostat is set properly, proceed to test it for defects and incorrect wiring. As always, refer to the recommendations made by the manufacturer during your troubleshooting.

The last component to check is the blower motor. It might be overheated or defective. Put the heat pump into the cooling mode and make the blower motor run. If the fan in the blower doesn't move, check for an open overload. Assuming that the motor is not overheated, you will have to replace it.

Ice

What would make the evaporator of your heat pump become covered in ice? A dirty air filter could be the cause of an ice build-up. Check the filter and replace it if necessary. Then, use the heat pump and see if the problem continues. In the meantime, check the air temperature in your home. If room temperature drops below 55 degrees, icing can occur on the evaporator.

The motor might be your problem. Check the motor to see that it is set on the speed specified by the manufacturer. It might be necessary to set the motor to a higher speed. Also, check the blower motor to see that it is not overheated or that it hasn't tripped off on overload.

There is one other possible cause for the problem, but it is rare to encounter it. If the water in the system is too cold, the evaporator could ice up. This is highly unlikely, but it has been known to happen.

Short cycles

What should you look for if your heat pump short cycles? The thermostat is the most likely cause of the problem. Assuming that you have installed the thermostat in a recommended location, check to see if the differential is set too close in the control. If necessary, adjust the heat anticipator to balance out the thermostat. Inspect all wiring to locate any loose wires, and pay attention to the control contactor to see if it is working properly.

If none of the early troubleshooting steps reveal the cause of the problem, check out the compressor and refrigerant charge. If the refrigerant charge is low, the compressor might be running hot. It is also possible that the compressor overload is kicking in and causing the short cycles.

Poor performance

Poor performance, in general, can be caused by a number of potential problems. If your heat pump is not producing enough warm or cool air, there will be many steps to take in finding the cause. Start with some of the easier aspects to check out and work your way through the many options that might be at fault for the poor performance.

Thermostat Where is your indoor thermostat located? Did you put it on an outside wall? Is it mounted where direct sunlight shines on it through a window? Improperly locating the thermostat can easily affect the performance of a heat pump. If you have put the thermostat in a location where extreme shifts in temperature occur, such as from direct sunlight, you should relocate it. The thermostat should be positioned on an inside wall, about five feet above the floor, in a location that is not subjected to hot or cold conditions that are different from the rest of the climate-controlled area.

Air flow Air flow is crucial to the successful operation of a heat pump. If return grilles are blocked, supply registers are closed, or dampers are malfunctioning, air flow will not be good and neither will the performance of the heat pump. Check for any reason that might cause insufficient air flow, like those just mentioned, and inspect all duct work to see that it is in place and not leaking conditioned air.

Refrigerant   Refrigerant is needed to keep a heat-pump system running effectively. If there is too much or too little refrigerant in the system, problems can occur. Check the refrigerant level in the system to ensure that it is at the level recommended by the manufacturer of your heat pump.

Water   Water is a key element in the operation of a water-source heat pump. If you are having problems with the performance of your heat pump, check the water pressure, temperature, and flow. If any of these aspects of the water are not what they should be, corrections will be necessary.

Reversing valve   The reversing valve in your heat-pump system is another key element in the system. If the reversing valve is defective, it can allow refrigerant to move from the discharge side of the system to the suction side of the compressor.

Contaminants   Contaminants in the refrigerant system can wreak havoc with a heat pump. Moisture is all it takes to mess up the system. Solid particles can also create problems, such as clogged strainers or filters. If you suspect that the refrigerant system has become contaminated, you should check strainers, filters, and capillary tubes for obstructions. It might very well be necessary to dehydrate and evacuate the whole system.

Operating pressure   Heat pumps have specific operating pressures that they should work within. Check the specifications provided by the manufacturer of your equipment to make sure your unit is running at the proper operating pressure.

Blowers   Blowers for heat pumps are sometimes wired backwards. If the capacitor leads are reversed in the motor, the fan can run backwards. This, of course, is not right, and reversing the capacitor leads will be necessary.

Compressors   Compressors are a major part of a heat pump. If the compressor is bad, the heat pump cannot operate properly. To check your compressor, you will need to identify its discharge pressure and suction pressure. If the discharge pressure is too low and the suction pressure is too high, there is a good chance the compressor is defective and will need to be replaced.

Size   The size of a heat pump is relevant to its ability to heat and cool appropriately. If a system is designed properly, there shouldn't be any problem related to the size of the heat pump. However, if a mistake was made in calculating loads and tonnage, the heat pump might be un-dersized. Minor mistakes in sizing can sometimes be made up for with insulation or other simple procedures. A major mistake in sizing could result in having to replace the heat pump with a larger unit.

Low pressure

If your heat pump is off because of the low-pressure cut-out control, the suction pressure might be too low during the cooling cycle. To verify this, you must inspect the blower, the water coil, and the filter.

If the heat pump is off because of the cut-out in a heating cycle, you should check the water flow and temperature in the system. Older heat pumps could have built up a scaling, usually due to hard water, in the water coil. This can contribute to restricted water flow and problems associated with it.

Two other possible causes for this problem are a defective low-pressure switch and a low refrigerant charge. In the case of the switch, you should check to see that it is not stuck in the open position. If the switch will not reset, it should be replaced. It is also possible that the switch is out of calibration.

In the case of a low refrigerant charge, you must add refrigerant to the system. There might be a leak in the system that caused the re-frigerant level to drop. Inspect for any leaks and repair them if found. Remember to evacuate the system before you recharge it if leaks are found or the system has been opened.

High pressure

Just as a heat pump can cut out due to low pressure, it can also cut out due to high pressure. The cause could be a defective high-pressure switch. Under such circumstances, you must test and evaluate the switch, just as was described for the low-pressure switch.

The refrigerant charge could be at fault, but this time it will be a matter of having too much refrigerant, rather than not enough. Check the refrigerant charge and bleed off any excess to bring the charge back within the system's designed level.

Other factors can contribute to a heat pump cutting off on the high-pressure control. If the system is shutting down during a cooling cycle, check the water flow and temperature. In older units, it is wise to inspect the water coil for a scale build-up.

When the unit is cutting off during a heating cycle, you should check air flow and temperature. It is also advisable to inspect the blower and filter. Again, if the system has some age on it, inspecting the coil for scale is a good idea.

Compressor won't run

When a compressor on a heat pump won't run, but the blower will, there are several potential problems to look for. Many of the checks involve working around electricity, so proceed cautiously, if at all.

The indoor thermostat for the system is the best place to begin your troubleshooting. Start by checking the thermostat settings. If they are all as they should be, check the wiring and calibration of the thermostat. Hopefully, you will identify the problem and be able to get on with more pleasant activities. If, however, the thermostat is not at fault, go over all of the wiring for the system and make sure the wiring has been done correctly and is not loose or broken. Also, check the voltage for the system to ensure that it is as is should be.

While on the subject of electrical checks, you should test the continuity of the compressor windings. This is done with an ohmmeter, but if you didn't already know that, you probably shouldn't be working around electricity. If your test proves the windings to be open, the compressor will have to be replaced.

Your unit could be on the blink due to the high or low pressure cut-out controls. This is easy to test for, if you are accustomed to working with controls. The first step in the process is to turn the thermostat to its off position. Wait five minutes and put the thermostat selector into the cooling mode. When this is done and the compressor comes on, you will know the unit was off due to either the high or low pressure cut-offs. Should the compressor fail to start, one of the pressure switches might be bad. This can be tested by jumping the switches on an individual basis. By testing the switches one at a time, you can determine which one of them is faulty.

Moving on through the troubleshooting steps, you can test the lockout relay. To do this, turn the power off. Wait a few minutes and then turn the power back on. If the rely was stuck in an open position, turning the power off should have closed it. If the relay did not reset itself, it will have to be replaced. While you are experimenting, check the capacitor. If it is defective, replace it.

The last few potential causes for your problem all deal with the compressor. Perhaps the compressor overload is tripped. To check for this, see if the compressor is hot. If it is, the overload will not reset itself until the compressor has cooled down. Assuming that the compressor is not hot, the overload might be defective. If the overload is external, it can be replaced. An internal overload that has gone bad will require the replacement of the compressor.

Sometimes a compressor will seize up. This can be tested by connecting an auxiliary capacitor in parallel with the run capacitor for a few moments. With this done, the compressor might start. If the compressor starts but falls back into the same problem, an auxiliary start kit might solve your problem. When a compressor will not start even with an auxiliary capacitor hook up, the compressor will normally have to be replaced.

The last thing to check is the grounding or burnout of the compressor. Check the internal windings to see if they are grounded to the compressor shell. If they are, expect to replace the compressor. When a compressor has burned out, a filter drier should be installed on the suction line.

This completes all the basic troubleshooting options for both air-source and water-source heat pumps. When you buy your heat pump, the manufacturer will provide a list of instructions for handling problems with your specific unit. Always read and follow those instructions before attempting any troubleshooting or repair work.

Troubleshooting procedures

Troubleshooting procedures vary from individual to individual. While every service mechanic might know what typical symptoms indicate in a troubleshooting scenario, different mechanics will go about their troubleshooting in various ways. Some are more efficient than others.

Because most readers of this book are not seasoned HVAC mechanics, finding out how to troubleshoot problems in the easiest way possible is especially important. An inexperienced troubleshooter can create more problems than initially existed.

Trial and error

All troubleshooting is a matter of trial and error, to some extent. Service mechanics must make assumptions and test their theories. This is certainly a trial-and-error procedure, but it is not a blind approach. Random trial-and-error troubleshooting can find problems, but it is not the most efficient way to troubleshoot a system.

Experienced service mechanics have a big advantage over homeowners when it comes to troubleshooting a heat pump. The mechanics have field experience, and experience is hard to beat when seeking solutions to problems. Inexperienced people have to rely on books and manuals to get them through their troubleshooting. As good as the printed page can be in helping someone accomplish a difficult task, there is no replacement for field experience. For those of you with little experience, here are some secrets of successful troubleshooting that I've learned over the last 20 years.

Keep it simple

The first rule of effective troubleshooting is to keep the process simple. Too many people overlook simple solutions because they are so consumed with tracking technical problems. Always start with the easiest solutions and work towards the difficult ones. If you're lucky, you'll solve your problem before you get to the technical stuff. To illustrate this, let's look at an example of how a knowledgeable homeowner might troubleshoot a heat pump if it will not run.

The homeowner in this example is a regular do-it-yourselfer, who likes to work around the house and has a fair amount of mechanical ability. The homeowner wakes up one morning and finds that the heat pump is not running. Determined to find the cause of the problem and correct it alone, he starts out on his troubleshooting campaign.

The homeowner is armed with the manufacturer's troubleshooting checklist and his tool box. Reviewing the list of possible causes for his heat pump not running, the homeowner sees suggestions similar to the following:

- Tripped circuit breaker
- Loose wiring connections
- Defective transformer
- High-pressure control open
- Overload contacts open in compressor

The aggressive homeowner rolls up his sleeves and gets to work. The first thing he checks is his circuit-breaker box. No problem there. He breaks out his electrical meter and digs into the wiring. There is no power. This confuses the homeowner, but he tried diligently for over an hour to figure out what the problem is. His efforts are futile. He gives in and calls a professional to check out the system.

When the professional arrives, he goes to the breaker box and checks out the circuit. The breaker is on and holding. Next, the service person checks the disconnect boxes, to find that the disconnect lever has been turned off. The mechanic turns the disconnect on, and suddenly the system runs. The homeowner, who has observed this simple troubleshooting process, shrinks into the corner and wonders why he never thought to check the disconnect.

It is a mystery as to how the disconnect got turned off, but there is no question that the cause of the heat pump malfunction was simply a lack of power. This homeowner was so intent on finding a complex problem with his heat pump that he failed to look for the obvious.

This is a prime example of how a simple problem can get out of hand. An experienced service mechanic will start at a logical point, such as the breaker box, and work methodically towards the cause of the problem. No step will be overlooked, and every conceivable cause will be investigated along the way.

Here is a real-life situation to further illustrate this. I was once called to a house for a unit that was not running. When I arrived, the homeowner proudly explained all the troubleshooting steps that he had already taken. I started at the beginning anyway; the problem was a blown fuse. The homeowner had already replaced the fuse himself, but he replaced it with a defective fuse. He assumed that because the fuse had been replaced, it had to be good. Rule number two is to never assume anything when you are troubleshooting, always prove your theory.

Work in a planned order

Your troubleshooting will be far more effective if you work in a planned order. Look over the possibilities that might be causing your problem. Group them together. For example, if you have three potential causes related to the compressor, put those three items in one group. Place all of the possibilities surrounding the thermostat into another group, and so on.

Once you have all of your options grouped together, assess them for ease of elimination. For example, it is very simple to check the settings on the thermostat. It is also easy to inspect the floor registers to see that they are open. Perform these easy tasks first, they might be all that is required to get your heat pump back up and running.

By having your troubleshooting tasks grouped together, you won't waste time running from one part of your system to another and back again. All of you work can be done efficiently and effectively.

Keep notes

Keep notes of your work as you go along. If you are involved in a troubleshooting phase that encompasses a number of steps, the notes can become invaluable. They will enable you to know what steps have already been completed and what the results of those tests were. This can be very helpful as you progress through the troubleshooting progress, and the notes might come in handy for future problems. You will have some benchmark numbers to draw from, and you will know what the system was doing, or in some cases not doing, at particular times and under certain circumstances.

Be careful

Be careful not to create problems that don't exist. If you don't know what you're doing, it is easy to experiment and cause more trouble than you already have. For example, suppose you suspected the wiring to your blower was reversed. You might switch the wires around to see what would happen. If that maneuver didn't solve your problem, you would proceed with the troubleshooting, but would you remember that you had switched the wires? If you kept good notes you would. Should you forget about switching the wires, you might be creating a new problem. If the wires were not backwards to begin with, they are now.

Know when to ask for help

It is important that you know when to ask for help. It's great to be handy and to be able to fix your own problems, but guessing about what to do with a troublesome heat pump can be dangerous. You could cause damage to your equipment, and you could get yourself hurt.

Much of the troubleshooting done with heat pumps requires working with electricity. Anytime inexperienced hands dabble in electrical wiring, trouble can arise quickly. Improper wiring can keep your heat pump from running. It can damage components of the heat pumps, and it can kill you.

Working with refrigerants is not something most people do on a regular basis. If you don't have the equipment and the skills to properly dehydrate and evacuate a refrigerant line, you could spend days trying to solve a problem that a professional could take care of in a simple service call.

There are a lot of sensitive controls and components involved with a heat pump system. These items will not tolerate a lot of abuse. If you don't understand how the controls work or how you should go about working with them, call in a professional. There is no shame in knowing when to call for professional assistance.

See Tables 17-1 through 17-28, and Figs. 17-1 through 17-5.

Table 17-1 Heat pump will not defrost completely

Check refrigerant charge
Check defrost circuitry
Check defrost controls
Check compressor valves

Table 17-2 Auxiliary heat will not cut off after defrost cycle

Check for a defective defrost circuitry
Check for a defective defrost component
Check thermostat
Check emergency heat relay

Table 17-3 Defrost cycle will not stop

Check for defective defrost controls
Check for a defective temperature bulb
Check the refrigerant charge
Check for a dirty indoor coil
Check for dirty air filters
Check for a defective expansion valve

Table 17-4 Heat pump provides heat when it should be cooling

Check for defective reversing valve
Check for defective electric resistance elements

Table 17-5 Heat pump is not heating properly and is indicating a low suction pressure

Check for a defective expansion valve
Check for a defective outdoor fan or motor
Check refrigerant charge
Check for restricted tubes at outdoor coil
Check for dirty outdoor coil

Table 17-6 Heat pump is not heating properly and is indicating a high liquid pressure

Check for restricted air flow
Check refrigerant charge to see if it is too high
Check for dirty filters
Check for a dirty coil
Check the blower motor
Check for noncondensibles in the system

Table 17-7 Heat pump is not cooling properly and its operating pressures are not normal

Check for problems in the electrical system
Confirm that the heat pump is of a proper size
Check for air leaks

Table 17-8 Heat pump is not cooling properly and is indicating a high suction pressure

Check for defective compressor valves
Check for defective reversing valve
Check refrigerant charge to see if it is too high
Check for noncondensibles in the system

Table 17-9 Heat pump will not defrost

Check for closed relays on the reversing valve
Check for closed defrost relays
Check for obstruction in the pressure tube

Table 17-10 Heat pump is not heating properly and is indicating a low liquid pressure

Check for a leak in the system
Check for a defective check valve
Check refrigerant charge to see if it is too low
Check for a defective expansion valve
Check for defective compressor valves
Check for restrictions in the liquid line

Table 17-11 Heat pump is not heating properly and its operating pressures are not normal

Check outdoor thermometer for improper settings
Check for restricted air flow

Table 17-12 Heat pump compressor cycles off in the defrost cycle

Check for a defective check valve
Check for a defective expansion valve
Check for a dirty indoor coil
Check for a bad reversing valve
Check for a dirty filter

Table 17-13 Heat pump is not cooling properly and is indicating a high liquid pressure

Check for a dirty outdoor coil
Check to see that outdoor fan is operating properly
Check refrigerant charge to see if it is too high
Check for noncondensibles in the system
Check for restrictions in the liquid line

Table 17-14 Heat pump is not heating properly and is indicating a high suction pressure

Check for a defective check valve
Check for a defective reversing valve
Check refrigerant charge to see if it is too high
Check for defective compressor valves
Check for noncondensibles in the system

Table 17-15 Heat pump will not run

Check fuses and circuit breakers
Check for faulty wiring
Check thermostat setting, location, and calibration
Check compressor overloads to see if contacts are open
Check to see if high-pressure control is open
Check for defective transformer

Table 17-16 Heat pump is not cooling properly and is indicating a low suction pressure

Check for defective compressor valves
Check for leaks
Check refrigerant charge to see if it is too low
Check for dirty air filters
Check for dirty indoor coil
Check for restrictions in the liquid line coupling
Check for defective check or expansion valve

Table 17-17 Heat pump is off on low-pressure cut-out control

Check refrigerant charge
Check for low suction pressure on cooling cycle
Check for low suction pressure on heating cycle
Check for defective low-pressure switch

Table 17-18 Heat pump runs but compressor doesn't

Check pressure controls
Check electrical wiring
Check thermostat
Check lockout relay
Check for a seized compressor
Check voltage supply
Check to see if the windings of the compressor are open

Table 17-18 Continued

Check to see if the compressor is overloaded
Check for a defective capacitor
Check to see if the compressor motor is grounded

**Table 17-19 Heat pump produces
inadequate heating or cooling**

Check for leaks
Check to be sure the heat pump is not too small
Check to see that sufficient water pressure is available
Check to see that thermostat is not improperly located
Check for inadequate air flow
Check for a lack of refrigerant
Check for a defective compressor
Confirm that the blower is not blowing in reverse
Check for a defective reversing valve
Check operating pressure
Check refrigerant system

Table 17-20 Heat pump will not run when calling for heat

Check for a clogged air filter
Check for an improperly set thermostat
Check for a defective thermostat
Check for a defective blower motor
Check for problems with the electrical wiring

Table 17-21 Heat pump is not cooling properly

Check for a clogged air filter
Check for a restricted water flow through coil
Check for refrigerant leaks
Check for a defective compressor

Table 17-22 Heat pump will produce only heat

Check for a defective reversing valve

Table 17-23 Heat pump is icing up

Check for a clogged air filter
Check to be sure the motor is set at the right speed
Check to see if the water temperature is too low
See if the thermostat is set for a very low temperature

Table 17-24 Heat pump is short cycling

Check for a compressor overload
Check electrical wiring
Check thermostat for location and defects
Check lockout relay
Check high-pressure cut-out
Check discharge pressure to see if it is too high
Check refrigerant charge
Check for inadequate water flow
Check for excessive air flow
Check for defective high-pressure switch

Table 17-25 Heat pump will not run

Check fuses and circuit breakers
Check for broken or loose electrical wires
Check for a possible low voltage supply or circuit
Check thermostat

Table 17-26 Troubleshooting heating Friedrich Air Conditioning Co.

PROBLEM	POSSIBLE CAUSE	TO CORRECT
No heating – fan operates.	Thermostat setting.	Set thermostat to a warmer position.
	Defective thermostat.	Replace – do not attempt to adjust.
	Compressor not operating.	Check compressor wiring. Check for open internal or external overload. Check wiring.
	Defective system switch.	Test system switch.

PROBLEM	POSSIBLE CAUSE	TO CORRECT
Insufficient heating.	Restricted filter.	Clean as recommended in Owner's Manual.
	Outdoor thermostat. (Applicable models.)	Check if outdoor thermostat is energizing the heating element at its predetermined temperature setting.
	Fresh air or exhaust door open.	Check control setting.

PROBLEM	POSSIBLE CAUSE	TO CORRECT
Fan operates in "constant" position, but not in "automatic", (MoneySaver).	Inoperative system switch.	Check continuity of switch.
	Incorrect wiring.	Check applicable wiring diagram.

Table 17-26 Continued

PROBLEM	POSSIBLE CAUSE	TO CORRECT
Temperature varies from comfortable to overly warm.	Defective thermostat.	Incorrect differential setting. Replace thermostat.
	Heat anticipator (resistor) shorted. (Applicable models.)	Check voltage to resistor. If voltage okay, remove resistor from thermostat bulb block. With current on, feel resistor for warmth. If no heat can be felt, replace anticipator.

PROBLEM	POSSIBLE CAUSE	TO CORRECT
Room temperature uneven. (Heating cycle)	Heat anticipator (resistor) shorted. (Applicable models.)	Disconnect power to unit. Remove resistor from thermostat bulb block. Plug in unit and allow to operate. Feel resistor for heat. If no heat is felt, replace resistor.
	Wide differential – partial loss of thermostat bulb charge.	Replace thermostat and check.
	Incorrect wiring.	Refer to appropriate wiring diagram. Resistor is energized during the "ON" cycle of compressor or fan.

PROBLEM	POSSIBLE CAUSE	TO CORRECT
Unit will not defrost.	Incorrect wiring.	Refer to appropriate wiring diagram.
	Defrost control timer motor not advancing.	Check for voltage at "TM" and "TM1" on timer. If voltage, replace control.
	Defrost control out of calibration.	If outside coil temperature is 25°F or below, and preselected time limit has elapsed, replace defrost control.
	Defrost control contacts stuck.	If contacts remain closed between terminals "2" and "3" of the defrost control after preselected time interval has passed, replace control.
	Defrost control bulb removed from coil, or not making good coil contact.	Reinstall and be assured that good bulb to coil contact is made.

PROBLEM	POSSIBLE CAUSE	TO CORRECT
Unit does not heat adequately.	Outdoor thermostat does not cut off compressor at the preselected temperature and bring on heating element.	Defective thermostat – replace.
	Fresh air or exhaust door open.	Check if operating properly. Instruct customer on proper use of control.
	Dirty filter.	Clean as recommended in Owner's Manual.
	Unit undersized.	Check heat rise across coil. Refer to performance data sheet on heat rise at various outdoor ambients. If heat rise is satisfactory, check if insulation can be added to attic or walls.

PROBLEM	POSSIBLE CAUSE	TO CORRECT
Unit cools when heat is called for.	Incorrect wiring.	Refer to applicable wiring diagram.
	Defective solenoid coil.	Check for continuity of coil.
	Reversing valve fails to shift.	Block condenser coil and switch unit to cooling. Allow pressure to build up in system, then switch to heating. If valve fails to shift, replace valve.
	Inoperative system switch.	Check for continuity of system switch.

(Cooling/Electric Models)

PROBLEM	POSSIBLE CAUSE	TO CORRECT
Fan operates – heating element does not come on.	Heater relay or contactor coil open.	Check continuity of coil.
	Heater relay or contactor stuck open, pitted or burned.	Inspect, test continuity with ohmmeter.
	High limit control open.	Check continuity – if open, replace.
	Open thermal fuse.	Check continuity. Check reason for failure.
	Open or shorted element.	Check voltage across heater terminals. Check amperage draw of heater.
	Loose connections.	Tighten all terminals.

PROBLEM	POSSIBLE CAUSE	TO CORRECT
Heating inadequate.	Restricted filter.	Clean as recommended in Owner's Manual.
	Cycling high limit control.	Control is set to open at 155°F ± 5°F and close at 130°F ± 8°F. If cycling prematurely, replace control.
	Exhaust or fresh air door open.	Check position of fresh air door control slide. Adjust cable if door does not close properly.

PROBLEM	POSSIBLE CAUSE	TO CORRECT
Fan operates in "Constant" position, but not in "Automatic" (MoneySaver).	Fan relay contacts open.	Check continuity of fan relay. NOTE: Some models have the fan relay energized during the heating cycle while others do not.
	Inoperative system switch.	Check continuity between terminals "L2" and "3" of the system switch.
	Loose connection.	Check connections on system switch and fan relay.

PROBLEM	POSSIBLE CAUSE	TO CORRECT
Long "off" and "on" cycles.	Heat anticipator (resistor) shorted.	Disconnect power to unit. Remove resistor from thermostat bulb block. Plug in unit and allow to operate. Feel resistor for heat. If no heat is felt, replace resistor.
	Defective thermostat.	Replace thermostat and check operation.

PROBLEM	POSSIBLE CAUSE	TO CORRECT
Cooling adequate – heating insufficient.	Heating capillary tube partially restricted.	Check for partially starved outer coil. Replace heating capillary tube.
	Check valve leaking internally.	Switch unit several times from heating to cooling. Check temperature rise across coil. Refer to specification sheet for correct temperature rise.
	Reversing valve failing to shift completely – bypassing hot gas.	Deenergize solenoid coil, raise head pressure, energize solenoid to break loose. If valve fails to make complete shift, replace valve.

Table 17-26 Continued

PROBLEM	POSSIBLE CAUSE	TO CORRECT
Compressor will not turn off and operate on heating element only during low outside ambients.	Outdoor thermostat. (Applicable models.)	Refer to the heating data on applicable models for the preselected temperature the compressor shuts off and the electric element is energized.

PROBLEM	POSSIBLE CAUSE	TO CORRECT
Compressor shuts off on outdoor thermostat but element does not heat.	Fuse link.	Check fuse link for continuity. If defective, replace.
	Heating element shorted.	Check amperage draw of element. If no amperage, replace.
	Incorrect wiring.	Check voltage to element. If voltage is okay, check wiring.
	Heat relay or heater contactor coil open	Defective coil. Test coil for continuity.

Table 17-27 Troubleshooting touch test chart. Friedrich Air Conditioning Co.

NOTES:
*Temperature of Valve Body.
* *Warmer than Valve Body.

VALVE OPERATING CONDITION	DISCHARGE TUBE from Compressor	SUCTION TUBE to Compressor	Tube to INSIDE COIL	Tube to OUTSIDE COIL	LEFT Pilot Capillary Tube	RIGHT Pilot Capillary Tube	POSSIBLE CAUSES	CORRECTIONS
	1	2	3	4	5	6		
Normal COOLING	Hot	Cool	Cool, as (2)	Hot, as (1)	*TVB	*TVB		
Normal HEATING	Hot	Cool	Hot, as (1)	Cool, as (2)	*TVB	*TVB		
							MALFUNCTION OF VALVE	
Valve will not shift from cool to heat	Check electrical circuit and coil.						No voltage to coil.	Repair electrical circuit.
							Defective coil.	Replace coil.
	Check refrigeration charge.						Low charge.	Repair leak, recharge system.
							Pressure differential too high.	Recheck system.
	Hot	Cool	Cool, as (2)	Hot, as (1)	*TVB	Hot	Pilot valve okay. Dirt in one bleeder hole.	Deenergize solenoid, raise head pressure, reenergize solenoid to break dirt loose. If unsuccessful, remove valve, wash out. Check on air before installing. If no movement, replace valve, add strainer to discharge tube, mount valve horizontally.
							Piston cup leak.	Stop unit. After pressures equalize, restart with solenoid energized. If valve shifts, reattempt with compressor running. If still no shift, replace valve.
Valve will not shift from cool to heat	Hot	Cool	Cool, as (2)	Hot, as (1)	*TVB	*TVB	Clogged pilot tubes.	Raise head pressure, operate solenoid to free. If still no shift, replace valve.
	Hot	Cool	Cool, as (2)	Hot, as (1)	Hot	Hot	Both ports of pilot open. (Back seat port did not close.)	Raise head pressure, operate solenoid to free partially clogged port. If still no shift, replace valve.
	Warm	Cool	Cool, as (2)	Warm, as (1)	*TVB	Warm	Defective compressor.	
Starts to shift but does not complete reversal	Hot	Warm	Warm	Hot	*TVB	Hot	Not enough pressure differential at start of stroke or not enough flow to maintain pressure differential.	Check unit for correct operating pressures and charge. Raise head pressure. If no shift, use valve with smaller port.
							Body damage.	Replace valve.
	Hot	Warm	Warm	Hot	Hot	Hot	Both ports of pilot open.	Raise head pressure, operate solenoid. If no shift, replace valve.
	Hot	Hot	Hot	Hot	*TVB	Hot	Body damage.	Replace valve.
							Valve hung up at mid-stroke. Pumping volume of compressor not sufficient to maintain reversal.	Raise head pressure, operate solenoid. If no shift, use valve with smaller ports.
	Hot	Hot	Hot	Hot	Hot	Hot	Both ports of pilot open.	Raise head pressure, operate solenoid. If no shift, replace valve.
Apparent leak in heating	Hot	Cool	Hot, as (1)	Cool, as (2)	*TVB	*TVB	Piston needle on end of slide leaking.	Operate valve several times, then recheck. If excessive leak, replace valve.
	Hot	Cool	Hot, as (1)	Cool, as (2)	** WVB	** WVB	Pilot needle and piston needle leaking.	Operate valve several times, then recheck. If excessive leak, replace valve.
Will not shift from heat to cool	Hot	Cool	Hot, as (1)	Cool, as (2)	*TVB	*TVB	Pressure differential too high.	Stop unit. Will reverse during equalization period. Recheck system.
							Clogged pilot tube.	Raise head pressure, operate solenoid to free dirt. If still no shift, replace valve.
	Hot	Cool	Hot, as (1)	Cool, as (2)	Hot	*TVB	Dirt in bleeder hole.	Raise head pressure, operate solenoid. Remove valve and wash out. Check on air before reinstalling. If no movement, replace valve. Add strainer to discharge tube. Mount valve horizontally.
	Hot	Cool	Hot, as (1)	Cool, as (2)	Hot	*TVB	Piston cup leak.	Stop unit, after pressures equalize, restart with solenoid deenergized. If valve shifts, reattempt with compressor running. If it still will not reverse while running, replace valve.

VALVE OPERATING CONDITION	DISCHARGE TUBE from Compressor	SUCTION TUBE to Compressor	Tube to INSIDE COIL	Tube to OUTSIDE COIL	LEFT Pilot Capillary Tube	RIGHT Pilot Capillary Tube	NOTES: *Temperature of Valve Body. * *Warmer than Valve Body.	
	1	2	3	4	5	6	POSSIBLE CAUSES	CORRECTIONS
Normal COOLING	Hot	Cool	Cool, as (2)	Hot, as (1)	*TVB	*TVB		
Normal HEATING	Hot	Cool	Hot, as (1)	Cool, as (2)	*TVB	*TVB		
	Hot	Cool	Hot, as (1)	Cool, as (2)	Hot	Hot	Defective pilot.	Replace valve.
	Warm	Cool	Warm, as (1)	Cool, as (2)	Warm	*TVB	Defective compressor.	

Valve operated satisfactorily PRIOR to compressor motor burnout — caused by dirt and small gassy particles inside the valve. TO CORRECT: Remove valve, thoroughly wash it out. Check on air before reinstalling, or replace valve. Add strainer and filter-drier to discharge tube between valve and compressor.

Table 17-28 Troubleshooting Cooling. Friedrich Air Conditioning Co.

PROBLEM	POSSIBLE CAUSE	TO CORRECT
Compressor does no run.	Low voltage.	Check for voltage at compressor. 115 volt and 230 volt units will operate at 10% voltage variance.
	Thermostat not set cold enough or inoperative.	Set thermostat to coldest position. Test thermostat and replace if inoperative.
	Compressor hums but cuts off on overload.	Hard start compressor. Direct test compressor. If compressor starts, add starting components.
	Open or shorted compressor windings.	Check for continuity and resistance.
	Open overload.	Test overload protector and replace if inoperative.
	Open capacitor.	Test capacitor and replace if inoperative.
	Inoperative system switch.	Test for continuity in all positions. Replace if inoperative.
	Broken, loose or incorrect wiring.	Refer to appropriate wiring diagram to check wiring.

PROBLEM	POSSIBLE CAUSE	TO CORRECT
Fan motor does not run.	Inoperative system switch.	Test switch and replace if inoperative.
	Broken, loose or incorrect wiring.	Refer to applicable wiring diagram.
	Open capacitor.	Test capacitor and replace if inoperative.
	Fan speed switch open.	Test switch and replace if inoperative.
	Inoperative fan motor.	Test fan motor and replace if inoperative (be sure internal overload has had time to reset.)

Table 17-28 Continued

PROBLEM	POSSIBLE CAUSE	TO CORRECT
	Undersized unit.	Refer to Sizing Charts.
	Thermostat open or inoperative.	Set to coldest position. Test thermostat and replace if necessary.
	Dirty filter.	Clean as recommended in Owner's Manual.
Does not cool, or cools only slightly.	Dirty or plugged condenser or evaporator coil.	Use steam or detergents to clean.
	Poor air circulation in area being cooled.	Adjust discharge air louvers. Use high fan speed.
	Fresh air or exhaust air door open on applicable models.	Close doors. Instruct customer on use of this feature.
	Low capacity – undercharge.	Clean for leak and make repair.
	Compressor not pumping properly.	Check amperage draw against nameplate. If not conclusive, make pressure test.

PROBLEM	POSSIBLE CAUSE	TO CORRECT
	Fuse blown or circuit tripped.	Replace fuse, reset breaker. If repeats, check fuse or breaker size. Check for shorts in unit wiring and components.
Unit does not run.	Power cord not plugged in.	
	System switch in "Off" position.	Set switch correctly.
	Inoperative system switch.	Test for continuity in each switch position.
	Loose or disconnected wiring at switch or other components.	Check wiring and connections. Reconnect per wiring diagram.

PROBLEM	POSSIBLE CAUSE	TO CORRECT
	Dirty filter.	Clean as recommended in Owner's Manual.
	Restricted air flow.	Check for dirty or obstructed coil – clean as required.
Evaporator coil freezes up.	Inoperative thermostat.	Test for shorted thermostat or stuck contacts.
	Short of refrigerant.	De-ice coil and check for leak.
	Inoperative fan motor.	Test fan motor and replace if inoperative.
	Partially restricted capillary.	De-ice coil. Check temperature differential across coil. Touch test coil return bends for same temperature. Test for low running current.

PROBLEM	POSSIBLE CAUSE	TO CORRECT
	Excessive heat load.	Unit undersized. Test cooling performance of unit. Replace with larger unit.
	Restriction in line.	Check for partially iced coil. Check temperature split across coil.
Compressor runs continually. Does not cycle off.	Refrigerant leak.	Check for oil at silver soldered connections. Check for partially iced coil. Check split across coil. Check for low running amperage.
	Thermostat contacts stuck.	Check operation of thermostat. Replace if contacts remain closed
	Thermostat incorrectly wired.	Refer to appropriate wiring diagram.

PROBLEM	POSSIBLE CAUSE	TO CORRECT
Thermostat does not turn unit off.	Thermostat contacts stuck.	Replace thermostat.
	Thermostat set at coldest point.	Turn to higher temperature setting to see if unit cycles off.
	Incorrect wiring.	Refer to appropriate wiring diagram.
	Unit undersized for area to be cooled.	Refer to Sizing Chart.

PROBLEM	POSSIBLE CAUSE	TO CORRECT
Compressor attempts to start, or runs for short periods only. Cycles on overload.	Overload inoperative. Opens too soon.	Check operation of unit. Replace overload if system operation is satisfactory.
	Compressor attempts to start before system pressures are equalized.	Allow a minimum of two (2) minutes for pressures to equalize before attempting to restart. Instruct customer of waiting period.
	Low or fluctuating voltage.	Check voltage with unit operating. Check for other appliances on circuit. Air conditioner should be on separate circuit for proper voltage and fused separately.
	Incorrect wiring.	Refer to appropriate wiring diagram.
	Shorted or incorrect capacitor.	Check by substituting a known good capacitor of correct rating.
	Restricted or low air flow through condenser coil.	Check for proper fan speed or blocked condenser.
	Compressor running abnormally hot.	Check for kinked discharge line or restricted condenser. Check amperage.

PROBLEM	POSSIBLE CAUSE	TO CORRECT
Thermostat does not turn unit on.	Loss of charge in thermostat bulb.	Place jumper across thermostat terminals to check if unit operates. If unit operates, replace thermostat.
	Loose or broken parts in thermostat.	Check as above.
	Incorrect wiring.	Refer to appropriate wiring diagram.

PROBLEM	POSSIBLE CAUSE	TO CORRECT
Noisy operation.	Poorly installed unit.	Refer to Installation Instructions for proper installation.
	Fan blade striking chassis.	Reposition – adjust motor mount.
	Compressor vibrating.	Check that compressor grommets have not deteriorated. Check that compressor mounting parts are not missing.
	Improperly mounted or loose cabinet parts.	Check assembly and parts for looseness rubbing and rattling.

PROBLEM	POSSIBLE CAUSE	TO CORRECT
Water leaks into room.	Evaporator drain pan overflowing.	Clean obstructed drain trough.
	Condensation forming on base pan.	Evaporator drain pan broken or cracked. Reseal or replace.
	Poor installation resulting in rain entering room.	Check installation instructions. Reseal as required.
	Condensation on discharge grilles.	Dirty evaporator coil – clean. Very high humidity level.

Table 17-28 Continued

PROBLEM	POSSIBLE CAUSE	TO CORRECT
Thermostat short cycles.	Thermostat differential too narrow.	Replace thermostat.
	Plenum gasket not sealing, allowing discharge air to short cycle thermostat.	Check gasket. Reposition or replace.
	Restricted coil or dirty filter.	Clean and advise customer of periodic cleaning of filter.
	Tubular insulation missing from top of thermostat bulb.	Replace tubular insulation on bulb. (Applicable models.)
	Thermostat bulb touching thermostat bulb support bracket.	Adjust bulb bracket. (Applicable models.)

PROBLEM	POSSIBLE CAUSE	TO CORRECT
Prolonged off cycles (automatic operation).	Anticipator (resistor) wire disconnected at thermostat or system switch.	Refer to appropriate wiring diagram.
	Anticipator (resistor shorted or open). (Applicable models.)	Disconnect plug from outlet. Remove resistor from bracket. Insert plug and depress "Cool" and "Fan-Auto (MoneySaver)" buttons. Place thermostat to warmest setting. Feel resistor for temperature. If no heat, replace resistor.
	Partial loss of charge in thermostat bulb causing a wide differential.	Replace thermostat.

PROBLEM	POSSIBLE CAUSE	TO CORRECT
Switches from cooling to heating.	Thermostat sticking.	Change room thermostat.
	Incorrect wiring.	Refer to appropriate wiring diagram.

PROBLEM	POSSIBLE CAUSE	TO CORRECT
Outside water leaks.	Evaporator drain pan cracked or obstructed.	Repair, clean or replace as required.
	Water in compressor area.	Detach shroud from pan and coil. Clean and remove old sealer. Reseal, reinstall and check.
	Obstructed condenser coil.	Steam clean.
	Fan blade and slinger ring improperly positioned.	Adjust fan blade to $\frac{1}{2}$" clearance from condenser coil.

PROBLEM	POSSIBLE CAUSE	TO CORRECT
High indoor humidity.	Insufficient air circulation in air conditioned area.	Adjust louvers for best possible air circulation.
	Oversized unit.	Operate in "Fan-Auto (MoneySaver)" position.
	Inadequate vapor barrier in building structure, particularly floors.	Advise customer.

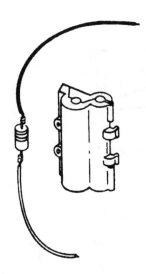

17-1
Thermostat block and resistor.
Friedrich Air Conditioning Co

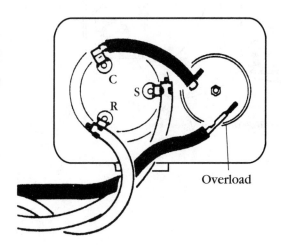

17-2
External overload. Friedrich
Air Conditioning Co

Overload

Line break
internal overload

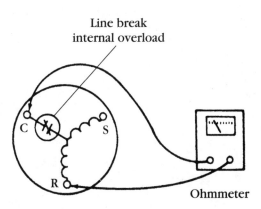

17-3
Internal overload. Friedrich Air
Conditioning Co.

Ohmmeter

17-4 *Compressor winding test.* Friedrich Air Conditioning Co.

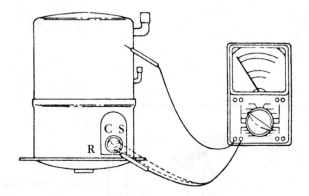

17-5 *Typical ground test.* Friedrich Air Conditioning Co.

18

Fix it fast
and fix it right

The thought of going through a cold winter night without heat in your home is not a pleasant one. Sweating through boiling summer days without air conditioning is no fun. There is never a good time for a heat pump to break down. Unfortunately, it seems that when they do quit working, they somehow pick the worst time to do it.

Not all heat-pump failures occur with old equipment. New heat pumps can quit working as quickly as they started. Unless you are willing to rely on highly paid professionals to keep you warm in the winter and cool in the summer, you must learn a bit about what makes a heat pump work and how to keep the unit running.

As a homeowner and amateur HVAC technician, you must understand your limitations. Professional HVAC mechanics spend years in both training and field experience to learn how to do the work that you are learning from this book and the information provided by the manufacturer of your heat pump. You cannot expect to have all the equipment and skill needed to overcome all the potential problems that can arise around a heat pump. Keep this in mind as you tackle tough jobs.

I want to stress that much of the work discussed at this point will be beyond the abilities of most people. There are many instances where only a professional will be able to perform the work near perfection. As we go along, I will mention my feelings on when an average homeowner should defer to a professional for the needed repair. Factor in your own common sense; be aware of your own limitations. You will know if you are comfortable doing a job as it is described. If you're not sure you understand what to do or when to do it, call in professional help. In all cases, heed the manufacturer's recommendations provided with your heat-pump equipment.

Control boxes

All heat pumps have control boxes that house a majority of the electrical components used to keep the heat pump working. In split systems, which are used for most residential requirements, the control boxes are located in the outdoor unit. The exact location will vary from manufacturer to manufacturer, but the control box should be somewhere in the outside unit.

When you look inside a control box, you are likely to find a number of components. For example, capacitors, defrost controls, contactors, start-assist devices, electrical circuits, and electrical components can all be found in control boxes. Each heat pump will be supplied with a wiring diagram for the equipment inside the control box, and these diagrams should be followed closely.

When you begin to work in a control box, you will be working around electricity. Avoid working with or around electricity unless you have experience in doing so. The risk of injury far outweighs the cost of hiring a professional service person to perform the work for you.

Capacitors

Capacitors are used in conjunction with motors to provide increased torque for starting the motor. An average control box might contain up to three capacitors. One of them is to help in starting the fan motor. Another is used in conjunction with the motor for the compressor. If there is a third capacitor, it works as a start capacitor.

When a capacitor is not functioning properly, it can have an adverse affect on the motor it is serving. To check a capacitor, the first thing you should do is look at it closely. Is the case of the capacitor swollen? Are any leaks present? If the capacitor is not bulging or leaking, you should proceed to test the component in a more technical manner. To do this, you must first have an ohmmeter.

Discharge the capacitor with 15,000-ohm and two-watt resistor. Then all wires should be removed. Your ohmmeter scale should be set to R × 100. At this time, place the ohmmeter probes across the terminals. What is the reading from your meter?

The meter should start at a low resistance value and move to a measurable resistance before it stops. When the indicator on the ohmmeter doesn't move at all, the capacitor is open and must be replaced. When an indicator that registers a zero reading and maintains it, the capacitor is shorted out and must be replaced.

There is another test that you can try on capacitors. This test is best done with the use of a capacitor tester. When such an instrument is not available, a substitution can be made with an ammeter and a voltmeter. To use an ammeter and a voltmeter, you must set the voltmeter in parallel with the charging circuit. The ammeter jaws should surround one leg of the charging circuit. With this done, energize the circuit long enough to get a meter reading.

Once you have your reading, compute the capacitance by multiplying the amps by 2650. Then divide that number by the volts. The numerical answer to the equation should fall within 10% of the rating on the case value. It can be either 10% above or below the case rating and still be all right.

Protector modules

You might find that your heat pump is equipped with a solid-state motor protector module. This module will be working in conjunction with the compressor. Because these modules work with the use on internal sensors that are buried in the motor windings of the compressor, you cannot condemn a module until both the module and the sensors are tested. This test begins with the module itself.

First, you must turn off the electrical power to the unit. Wait unit the compressor is cool, and then place a jumper wire between the terminals of the module. Apply electrical power and bring power to the control circuit by setting the unit to call for cool air. If the compressor operates, you might be on the right track in blaming the module for your problem, but you will have to test the sensors to be sure. If the compressor doesn't operate, there might be a problem in the low-voltage electrical circuit, but you can rule out the protector module.

When you want to test the sensors, you must remove the leads from the module. Be careful not to mix up the leads in a way that will make putting them back in their proper place difficult. You might want to wrap some masking tape around the leads and mark them so that you will know exactly where to reinstall them.

You will need an ohmmeter to test the sensors. The sensors are sensitive, and no more than five volts should be applied to them. Any additional voltage could cause damage to the sensors. The ohmmeter used to test the sensors must have a range of zero to 200,000 ohms.

When you test the sensor, the lack of a meter reading will indicate an open circuit. A zero reading on the meter points to a shorted or grounded circuit. Compare any other reading you obtain with the paperwork that came with your equipment. If the reading is within the acceptable range for the sensor, the protector module is bad and must be replaced. When you find that the sensors themselves are bad, the whole compressor will have to be replaced.

If your test indicates a need to replace the compressor, I would recommend that you call in a professional to render a second opinion. It is possible that you made a mistake in your testing. The purchase and installation of a new compressor that is not needed is an expensive on-the-job lesson to learn. By calling in a professional, you can obtain a concurrence on your conclusion before you invest in a new compressor.

Variable resistors

Variable resistors are found in some heat pumps. These devices take the place of a relay and capacitor combination. The purpose of the variable resistor is to give a little added punch of voltage when it is needed to get a motor started. If this resistor is bad, the motor might not be able to start.

Testing a variable resistor is not difficult. You need an ohmmeter to conduct the test. Turn off the electrical power to the unit, and disconnect the wires from the starting device. Place the probes of the ohmmeter over the terminals and take a reading. If you don't get any reading, the device is open and must be replaced. Any good meter reading indicates that the circuit is complete and need not be replaced.

Compressors

Most compressors are hermetically sealed. This means that work on them in the field is very limited in scope. Due to their design and construction, compressors are one part of a heat pump that just will not allow a lot of on-site tinkering. Unfortunately, the fact that little can be done with a compressor in the field means that compressors usually must be replaced rather than repaired. There are, however, some on-site tests you can run, if you are equipped to do so, that will indicate if a compressor must be replaced.

An overcharged refrigerant system can damage a compressor. If the refrigerant is charged to a point where it returns to the compressor, problems are going to occur. One of the tell-tale signs of such a condition is a noisy compressor. There is another way to tell if a compressor is bad or if it is the valves working with the compressor that need attention. This test, however, requires the use of a gauge manifold set. Because this is a tool you are not likely to have, it is probably best to call in a professional to run the test for you. But here are the steps involved.

The valves that work in connection with compressors to provide seals between the low and high pressure sides of the system are very important. If these valves are damaged or malfunctioning, the compressor is in trouble. Before a troubleshooter can know if a compressor problem is in the compressor or in the valves, a test must be conducted with the use of a gauge manifold set.

The test checks to see if suction pressure will pull down and if the discharge will build up. Assuming that the system has a proper refrigerant charge, the test will reveal the condition of the compressor. If the suction pressure will not pull down and the discharge will not build up, the compressor is most likely defective. You should watch the gauge pressures when the system is shut down. If they equalize quickly, the problem might be in the valves, rather than with the compressor.

What would make a compressor motor fail? If it becomes grounded from wires touching the motor stator or crankcase, it could fail. When a winding becomes open, the compressor can fail. And, when windings short together, failure is likely.

Off on overload

Some compressors are equipped with an internal line-break overload that can force the compressor to go off on overload. Not all compressors are equipped with these, and if yours is, it should be tested before you scrap the compressor. Check the paperwork from the heat-pump manufacturer to determine if this is a check you should make.

The line-break overload reads current and winding temperature. Under extreme conditions, the device will shut down the compressor. If you have an overload protector, you will need an ohmmeter to check it out.

To test the overload device, you should first turn off all power to the unit. If necessary, wait until the compressor has cooled before proceeding with the test. When the compressor is cool and the power is off, remove the wires from the compressor terminal. Your ohmmeter should be set at R × 1.

When you look in the terminal box, you see three terminals. One is marked with a "C." This indicates the common terminal. Another terminal is marked with the letter "S." This terminal is the start terminal. The last terminal is the run terminal, and it is marked with the letter "R."

Start your test with one probe of the ohmmeter on the common terminal and the other probe on the run terminal. Note the meter reading. Now, move the probe from the run terminal to the start terminal, and note the meter reading. A reading of low resistance means that you have ruled out the overload as being the problem. A reading of infinite resistance indicates that the overload might be open. To confirm this, position the probes of your meter on the run and start terminals. A reading of low resistance in this position verifies that the overload is open.

Check the windings

You can also check the windings of the motor. Turn your ohmmeter to the R × 100 scale, and then check to see if a winding is grounding out. To do this, place one of the ohmmeter probes on the compressor's discharge line. Move the other probe across the common, run, and start terminals. Keep a watchful eye on the meter as you touch the probe to the various terminals, and hope that you don't get a reading. If you do get a reading from any of the terminals, you've got a bad compressor.

There is another test that you should perform if the compressor has passed the previous tests. Set your ohmmeter to the R × 1 scale, and connect one of the probes to the common terminal. Touch the other probe to the run terminal and notice if there is any reading. Then, move the probe from the run terminal to the start terminal, and note the meter reading.

If you don't get any meter reading from the test, one of the windings is open. A zero reading or no reading indicates a short. Hopefully, you will get a low resistance reading, proving that the windings are okay. Check the specifications provided with your equipment to establish what readings you should obtain from this test.

Before you start the compressor, connect an ammeter to the common terminal. When you turn on the power, take note of the meter reading. Compare the reading with the Locked Rotor Amperage rating on the equipment information plate. The ammeter should draw at least the Locked Rotor Amperage amount when the motor starts, but then it should settle back down to a normal level. If the ammeter remains at a high reading, there is a problem with the compressor. It is also important to note that the hoses for the test gear must be connected to the proper refrigerant lines and that air from the hoses must be purged to prevent contamination of the refrigerant system.

Acid in a compressor

Acid in a compressor is destructive, and if a compressor burns out, there is a good chance that acid will mix with the oil in the compressor. If the acid is left alone, it will gradually eat away at the compressor, causing long-term damage. To avoid this, an in-line drier must be installed when a compressor has gone through a burn-out. Sometimes you can smell a burn-out when the system is opened, but the only way to tell for sure if acid has gotten into the oil is to test the oil. The damage-control sequence following a burn-out is one of the jobs that is probably best left to professionals. However, here are the basics of what should be done.

The first step is the removal of the bad compressor. Then, an R-22 cylinder should be attached to the suction line and used to purge the system of as much oil and solid contamination as possible. Once this is done, a new compressor can be put in place.

A clean-up filter/drier should then be installed on the common suction line. If an existing filter is in place, it should be removed and discarded. A filter/drier that utilizes a removable, replaceable core is preferable, but any clean-up filter/drier can be used. The filter/drier must be rated for one size larger than the tonnage of the unit it is serving.

The next step is to connect the suction and liquid lines to the compressor, and test the system for leaks. Then the entire system must be evacuated. When this is complete the system can be charged with refrigerant.

Turn the unit on, but don't change the mode of operation. In other words, if the system went down in the heating mode, start it in the heating mode. This will minimize the risk of spreading contaminants. Run the system for a few hours to make sure all the contaminants will be trapped. When you feel sure all of the contaminants are captured, remove the drier from the unit and recharge the system.

Evacuating a sealed system

Evacuating a sealed system is necessary if the system contains moisture. Moisture can invade the system any time the system is opened for repairs. To do this job professionally, a deep vacuum pump and a thermistor vacuum gauge are needed. A vacuum of 300 microns is pulled on the system and the pump is valved off. The system vacuum should not rise at a rate of more than 100 microns per hour. Sometimes the rate of micron rise is greater than the 100 microns per hour. This calls for more evacuation of the system until the systems vacuum will not rise faster than 100 microns per hour. As you might have guessed, this job is one that is usually done best by professionals.

Sometimes it is necessary to install a filter/drier on the common suction line of the system. Any existing filter/drier that was present when the system failed should be removed and discarded, so that a new one can be installed. A filter/drier with a replaceable core is the best type to install.

As mentioned, the system should be operated for a few hours in the same mode that it was in when the failure occurred. The filter/drier should be checked after a couple of hours for a drop in pressure. After complete removal of moisture and contaminants has been accomplished, the system can be recharged with refrigerant and put back into normal operation.

There is one other thing that should be noted. If you are evacuating a sealed system when the temperature outside is below 40 degrees, you must keep the outdoor unit warm. This can be done by placing a makeshift cover, like a tarp, over the outside unit and keeping a portable heater running near the compressor. Be careful, however, not to allow the portable heater to catch the tarp, or whatever cover is used, on fire.

Reversing valves

Reversing valves are often considered the heart of a heat-pump system. Without these valves, the heat pump would be unable to provide both warm and cool air. These valves are extremely important to the satisfactory operation of a heat pump.

Reversing valves control the flow of refrigerant in a heat-pump system. They are electrically controlled, and function on a pressure differential. Their sensitivity to pressure makes them rely on the system pressure to operate effectively. If the system pressure is out of kilter, the reversing valve will act up. This means that before you can pronounce a reversing valve to be defective, you must check not only the valve, but the system pressure as well. In fact, you should check the system pressure prior to testing the reversing valve. Refer to the manufacturer's specifications on operating pressures to determine if the system is running within the constraints of proper pressure loads.

Once you have established that the operating pressure of your system is at the proper level, you might begin to inspect the reversing valve. Start with a visual inspection of the valve body. Damage to the body of the valve can impede the slide assembly's movement on the inside of the valve. If you don't see any obvious problems, take your troubleshooting a step further.

Before you proceed with the reversing valve, disconnect the fan motor of the condenser and remove the fan blade. When this is done, you must provide power to the reversing valve. Its solenoid coil must be energized to work. What you are about to do is easier for someone who has worked with reversing valves in the past, because the test requires a certain feel for what the valve should do. Without field experience, you might have difficulty in judging the reactions of the test.

To test the solenoid coil and needle valve of the reversing valve, you must remove the lock nut that holds the coil in place. Remember that electrical power for the solenoid must be on during this test. After the lock nut is gone, slowly slide the coil off its stem. Do you feel any resistance in the coil movement? You should, if the coil is in operating condition. As you continue to slide the coil along the stem, you should hear a clicking sound. This is made by the plunger needle, and the sound indicates that the needle is responding to the magnetic field of the solenoid coil.

Assuming that you have felt resistance and heard a clicking sound, slowly slide the coil back into its intended place on the stem. Did you hear another clinking sound? If you did, it meant that the plunger was triggered by the coil, and this is good. If you felt no resistance or didn't hear both of the clicking sounds, there is an electrical problem to be checked out.

There is a wide array of problems possible with a reversing valve. So far, you've only touched the tip of the iceberg in determining if the valve is bad or not. To go further in your troubleshooting, you should have a chart that gives instructions for doing what is known as the "touch test." Without such a chart, you will have to hunt and peck to find out if the valve is bad.

When you look at a reversing valve, you will notice six tubes. One of these tubes is a discharge tube from the compressor. A second tube is a suction tube going to the compressor. There is also a tube that goes to the inside coil. A fourth coil leads to the outside coil. Then there are two other tubes, one for the left pilot capillary tube and another for the right pilot capillary tube. With the help of a chart provided by the manufacturer, you can perform a troubleshooting sequence by taking note of the temperature of each tube. The temperature is not measured with a thermometer. All that you have to do is touch the tubes and determine if they are hot, cool, or the same temperature as the body of the reversing valve.

During normal operation in the cooling mode, valve number 1, the discharge tube will be hot. Valve number 2, the suction tube, will be cool. The number 3 tube, the tube to the inside coil, will be cool. Tube number 4 will be hot. Tube number 5, the left pilot capillary tube, will feel the same as the reversing valve body. The right pilot capillary tube, tube number 6, will also be the same temperature as the valve body.

During normal operation in the heating mode, valve number 1, the discharge tube will be hot. Valve number 2, the suction tube, will be cool. The number 3 tube, the tube to the inside coil, will be hot. Tube number 4 will be cool. Tube number 5, the left pilot capillary tube, will feel the same as the reversing valve body. The right pilot capillary tube, tube number 6, will also be the same temperature as the valve body.

Now that you know how the various tubes should feel in both the heating and cooling mode, you can assess the condition of your reversing valve. If the temperatures of the tubes are not as they are described earlier, you have a problem with the valve. By noting the differences in the tube temperatures, you can rule out many potential problems and concentrate on the ones that are able to occur under the given circumstances.

If you have a chart from the manufacturer of your heat pump, you will notice that different forms of malfunctions are listed on it. In conjunction with the malfunctions are temperatures for the six tubes. By touching and judging the temperature of the each tube, you can move down your chart and eliminate some of the possible problems. Other problems will have to be investigated to greater lengths. To expand on this, let's go through some of the troubleshooting and repair procedures you might face with a reversing valve.

Will not shift

If the reversing valve will not shift from cool to heat, there are six likely causes of the problem. You could spend time checking all six of these potential causes, or you can use the knowledge you are gaining here to minimize your work and maximize your efforts.

One of the first things to check when the valve will not shift from cool to heat is the coil, as described earlier. If the coil is bad, it must be replaced.

Checking the refrigerant charge is the next logical step. If the refrigerant charge is low, it must be recharged. You should also inspect for any leaks that might have caused the refrigerant charge to drop. A pressure differential in the charge could be too high, so check the pressure ratings as well. Once you have done this, you are ready to test the temperatures of the tubes. Have a note pad with you to write your findings down; remembering the temperatures of each tube is critical in cutting down on the time it takes to troubleshoot the valve.

Let's start with worst possible outcome of your test. Notice the abbreviation of TVB; this stands for the same temperature as the valve body. Assume that you have just touched all of the tubes and came up with the following findings:

- Tube 1=warm
- Tube 2=cool
- Tube 3=cool
- Tube 4=warm
- Tube 5=TVB
- Tube 6=warm

If the tubes of your reversing valve have temperatures as indicated above, your compressor is probably bad. Now, suppose you had slightly different temperature recordings? Let's say that the temperatures were as follows:

- Tube 1=hot
- Tube 2=cool
- Tube 3=cool
- Tube 4=hot
- Tube 5=hot
- Tube 6=hot

These temperature readings would indicate that both parts of the pilot are open. The back seat port did not close for some reason; it might be clogged. With this being the case, you should raise the head pressure and operate the solenoid in an attempt to clear the obstruction. If you are still unable to make the reversing valve shift, it should be replaced.

As you can see, the temperature testing of the tubes can tell you a lot about your reversing valve. Here are some other scenarios to show you what the temperatures would indicate.

Assume you tested the temperature of the tubes and came up with the following results:

- Tube 1=hot
- Tube 2=cool
- Tube 3=cool
- Tube 4=hot
- Tube 5=TVB
- Tube 6=TVB

These readings would indicate clogged pilot tubes. To correct this problem, you would raise the head pressure and operate the solenoid to remove the obstructions. If this action was unsuccessful, you would have to replace the reversing valve.

Up to this point, each set of temperature readings has indicated a single problem. The troubleshooting process is not always that easy. Sometimes, there are more than one possibility for a problem associated with a given temperature reading. Here is an example of this.

You've just felt the six tubes associated with your reversing valve. The results are as follows:

- Tube 1=hot
- Tube 2=cool
- Tube 3=cool
- Tube 4=hot
- Tube 5=TVB
- Tube 6=hot

These readings indicate that you might have one of two problems. It will be up to you to determine which problem you have. Your readings prove that the pilot valve is okay, but there might be dirt in one of the bleeder holes. The other possibility is that a leak exists in the piston cup. The leak in the piston cup is easier to check for than dirt in the pilot is, so let's start with the piston cup.

Turn your heat pump off, and wait until the pressures equalize. Restart the unit, with the solenoid energized. If the valve shifts, try it again with the compressor running. If the valve won't shift with the compressor running, you must replace the reversing valve.

Assuming that the piston cup is not your problem, you must address the issue of dirt in the pilot. Kill the power to the solenoid, and raise the head pressure. Energize the solenoid in an attempt to break the dirt loose. If this doesn't work, remove the reversing valve and wash it out. Once the valve is clean, test it with air. If it won't work with air, the valve must be replaced. You should install a drier on the suction line and mount the valve horizontally.

Fails to shift completely

You might have a reversing valve that will start to shift but fails to shift completely. If you do, your temperature readings will look something like this:

- Tube 1=hot
- Tube 2=warm
- Tube 3=warm
- Tube 4=hot
- Tube 5=TVB
- Tube 6=hot

This type of reading means that either the body of the reversing valve is damaged, or there is not enough pressure differential at the start of the stroke or not enough flow to maintain the pressure differential. Obviously, you should inspect the valve body first. If it shows damage, replace it.

If the body of the valve is intact, you should check the operating pressure of the system. Also check the refrigerant charge. Raise the head pressure and see if the valve will shift. If it won't, install a valve with smaller ports.

Leak in heating

If you feel you have a leak in heating, there are two possible causes for the problem. Testing the temperature of the tubes will guide you to the problem that deserves attention. In the temperature readings you are about to be given, I've used another abbreviation. The letters WVB stand for a temperature that is warmer than the valve body. Here are the readings:

- Tube 1=hot
- Tube 2=cool
- Tube 3=hot
- Tube 4=cool
- Tube 5=TVB
- Tube 6=WVB

These readings indicate that the piston needle on the end of the slide is leaking. Try operating the valve several times to see if the leak stops, stays the same, or grows worse. If it is leaking badly, replace the valve.

The other possible cause of this broken would be having both the pilot needle and the piston needle leaking. The temperature readings would, however, be different if this were the case. Those readings would be as follows:

- Tube 1=hot
- Tube 2=cool
- Tube 3=hot
- Tube 4=cool
- Tube 5=WVB
- Tube 6=WVB

If this were the case, you should operate the valve several times and check on the condition of the leaks. If they are bad, the valve would have to be replaced.

Reversal not completed

If reversal is not completed after the reversing valve has started to shift, there are three possible causes for the problem. Again, temperature readings from the tubes will allow you to narrow the field of possibilities considerably. Let's look at the three different readings you might get under these circumstances and see what they will tell you. Here is the first set of readings:

- Tube 1=hot
- Tube 2=warm
- Tube 3=warm
- Tube 4=hot
- Tube 5=hot
- Tube 6=hot

The readings above indicate that both ports of the pilot are open. To correct this, raise the system's head pressure, and operate the solenoid. If the valve still will not shift, replace it. There is another combination of temperature readings that would indicate this same problem and repair procedure. Those readings are:

- Tube 1=hot
- Tube 2=hot
- Tube 3=hot
- Tube 4=hot
- Tube 5=hot
- Tube 6=hot

The remaining possibilities are that the body of the valve is damaged, or the compressor is not pumping enough to make the valve operate properly. If either of these are the case, your temperature readings should look like this:

- Tube 1=hot
- Tube 2=hot
- Tube 3=hot
- Tube 4=hot
- Tube 5=TVB
- Tube 6=hot

If the valve body is damaged, it must be replaced. When the valve is hanging up in midstroke, due to the compressor not pumping enough, you can try raising the head pressure to see if the valve will work. If it doesn't, replace it with a valve that has smaller ports.

From heat to cool

The last type of malfunction you might have to work with is when the reversing valve will not shift from heat to cool. There are six reasons why this might happen, but temperature readings from the valve tubes will cut the list down to size quickly.

Let's say that your temperature readings look like this:

- Tube 1=hot
- Tube 2=cool
- Tube 3=hot
- Tube 4=cool
- Tube 5=TVB
- Tube 6=TVB

The above readings indicate that the pressure differential is too high, or the pilot tube is obstructed. Start by raising the head pressure and operating the solenoid. This might free the obstruction from the pilot tube. If there is still no shift, and if the pressure differential is not too high, the valve must be replaced.

You can check to see if the pressure differential is a problem by shutting down the unit. The valve should reverse during the equalization period. If it doesn't, it's time to install a new reversing valve.

Now, assume the valve is the temperature readings are as follows:

- Tube 1=hot
- Tube 2=cool
- Tube 3=hot
- Tube 4=cool
- Tube 5=hot
- Tube 6=TVB

These readings point to dirt in a bleeder hole. Kill the power to the solenoid, and raise the head pressure. Energize the solenoid in an attempt to break the dirt loose. If this doesn't work, you must remove the reversing valve and wash it out. Once the valve is clean, test it with air. If it won't work with air, the valve must be replaced. You should install a drier on the suction line and mount the valve horizontally.

There is also a possibility that these same readings would indicate a leak in the piston cup. Turn your heat pump off, and wait until the pressures equalize. Restart the unit, with the solenoid energized. If the valve shifts, try it again with the compressor running. If the valve won't shift with the compressor running, replace the reversing valve.

What would the following temperature readings indicate?

- Tube 1=hot
- Tube 2=cool
- Tube 3=hot
- Tube 4=cool
- Tube 5=hot
- Tube 6=hot

These readings mean that you have a defective pilot and must replace the reversing valve. There is one last set of readings that you might get when the reversing valve will not shift from heat to cool. Those readings are going to indicate that the compressor is bad and must be replaced. The readings are as follows:

- Tube 1=warm
- Tube 2=cool
- Tube 3=warm
- Tube 4=cool
- Tube 5=warm
- Tube 6=TVB

Replacing a reversing valve

Replacing a reversing valve is not very difficult if you are able to evacuate and recharge the refrigerant system. There are, however, some aspects of the work that require patience, skill, and the observation of rules. Let's go through the sequence of what is involved with this process, and then you can decide whether it is beyond your capabilities or not.

The first step towards replacing a reversing valve is the removal of the solenoid coil. Once that is out of the way, you must break the solder joints that hold the valve in place. To loosen these joints, you are going to have to heat the copper tubing with a torch. But, you must not allow the body of the valve to get too hot. Even though you are going to trash the existing valve, too much heat on its body could introduce contaminants into the system.

To keep the valve body from overheating, wrap it in wet rags. The water in the rags will keep the body cool enough that no damage should be caused to the rest of the system. It might be necessary to pour additional water on the rags as you heat the pipes, so have some handy.

For obvious reasons, you won't want to pull the valve out of place with your bare hands. The copper tubing will be hot enough to brand you if it touches your skin. Use either heavy gloves or a pair of pliers to free the valve from the piping. Once the solder joints are loose, the valve can be removed. You must keep the joints hot enough to maintain liquidity of the solder, and you might have to twist the valve somewhat to get it free. If you've never done this before, you will probably have to make a few attempts before you are able to remove the valve.

When the old valve is out of the way, you should heat the ends of the copper tubing and wipe off all excess solder with a cloth. Remember that the tubing will be very hot, so wear thick gloves during this procedure. Once the ends of the tubing cool down, they should be sanded with a light-grit sandpaper. Now you are ready to install the new valve.

Just as the old valve wasn't allowed to overheat, the new valve must also be protected. Once the new valve has been placed on the tubing, wrap it with wet rags. Remember to protect the system and the valve tubes from any moisture that might attempt to get in.

The valve should be installed horizontally. It is also advisable to handle the valve carefully, to avoid any damage to the body. Check the valve to make sure it falls within an acceptable rating for your system, and solder it into place. Test your new installation for leaks, and then reinstall the solenoid coil.

Now it is time to add refrigerant to the system. Refer to the manufacturer's recommendations to determine the amount of refrigerant to add. With this done, you are ready to test the heat pump. Don't operate the reversing valve until you know the system is running properly. If the system is not charged properly, the reversing valve cannot work in the way it should.

When you are sure the system is working correctly, test the reversing valve. Put it through its cycles several times to check its operation. Assuming that it works well after a dozen or so cycles, pat yourself on the back for a job well done.

Superheat charging

When a professional technician is asked to charge a heat-pump system with refrigerant, the method most often used is called the *superheat-charging method*. This manner of charging a system is only done with the system in a cooling mode. As long as the system being charged is of a capillary-fed type, the superheat charging is very accurate.

To charge a system with refrigerant, you will need a few tools and the charts that were packed with your heat pump. To complete this chore, a thermometer is attached to the suction line where it enters the condensing unit. A suction gauge is attached to the suction line port of the condensing unit. The heat pump is then started and allowed to stabilize.

Once the system has stabilized, take readings for the pressure in the suction line and record them. Also test and record the temperature of the suction line. The last piece of the puzzle is the temperature of the outdoor air. With these three numbers, you can refer to the charts provided with your heat pump to see where you stand with your refrigerant charge.

Take your readings and plot them on the suction-line-temperature chart provided with your heat pump. As an example, the outside air temperature will be marked on the left side of the chart. The suction pressure will be labeled on the top of the chart. As you move from left to right, horizontally, to connect these two numbers with a corresponding number on the chart, you will see what the suction line temperature should be. If the temperature in the suction line of your system matches the recommendation on the chart, you're all set.

Normally, if the temperature reading for your system is lower than what is recommended, you will have to inspect the system to see if it is overcharged. If so, you must purge the refrigerant until the proper temperature is arrived at. A temperature in your system that is higher than what the chart recommends indicates the system is low on refrigerant and must be charged.

In addition to the chart that allows you to confirm the desired temperature of your suction line, there will also be a chart that provides recommendations for heating performance. You will have to use these charts to check the suction pressure of your system. To do this, you must take a temperature reading of both the outdoor air and the indoor air. With these two temperatures known, you can use the performance charts to establish desired discharge pressures and suction pressures. The paperwork that comes with your heat pump will explain how to use the charts.

Expansion valves

Many heat pump manufacturers provide thermostatic expansion valves to be used with their equipment. The troubleshooting and repair procedures involved with these valves are extensive.

If you decide to work with expansion valves, you should know how to measure and adjust the operating superheat. People who do not possess this knowledge and ability will be severely hampered in their troubleshooting and repair work with expansion valves.

Measuring and adjusting the operating superheat of a heat pump requires a temperature-pressure chart (this should be available from the manufacturer of your heat pump), a thermometer, and a suction gauge. The first step in the process is determining the suction pressure. This is done by attaching a gauge to the system and taking a reading. Once you have an accurate pressure reading, you can determine the saturation temperature by referring to the temperature-pressure chart supplied by the manufacturer of the heat pump.

The next step in measuring operating superheat involves taking the temperature of the suction gas. To do this, clean part of the suction line near the remote bulb location. Professionals would use a potentiometer to take the temperature reading, but it is possible to get a decent reading with a common thermometer. Tape the thermometer to the cleaned section of the suction line. It is important to protect the thermometer from ambient air temperature. You want a reading of the gas temperature, not the open-air temperature.

Once you have gotten an accurate reading of the suction gas temperature, you should subtract the saturation temperature that you determined from the temperature-pressure chart from the temperature of the suction gas. This will give you the superheat reading of the suction gas.

Assuming that your unit is equipped with an external adjustment valve, adjusting the operating superheat is no problem. If you remove the cap from the bottom of the valve, you should see an adjusting stem. This stem can be turned to increase or decrease the amount of refrigerant flowing through the valve. Turning the stem clockwise will decrease the flow. A counterclockwise rotation of the stem will increase the amount of refrigerant allowed to go by the valve. Increasing the flow of refrigerant will lower the superheat, and decreasing the flow of refrigerant will increase the superheat.

Not all expansion valves have external adjustments. If the adjustments on your valve are internal, the act of altering the superheat will be a bit more complicated. You will have to pump your unit down and disconnect the outlet line from the valve. Once you've done this, you can adjust the stem in the same way described in the last paragraph.

It's important to be familiar with some of the potential problems you might encounter that are involved with expansion valves. You might have an occasion when you have low suction pressure and high superheat. There are many reasons why this might happen. For example, the problem could be caused by the inlet pressure being too low from an excessive vertical lift, undersized liquid line, or excessive low condensing temperature. To correct this problem, you would have to increase the head pressure. If the problem was in the liquid line being too small, it would have to be replaced with a larger-diameter line. This, however, is only one of many possible causes and solutions for the problem; let's look at the rest of them.

Valve restriction

You might have a valve that is restricted by pressure drop through the evaporator. If this is the case, you should change to an expansion valve that has an external equalizer. But, let's say that your valve already has an external equalizer, and it is what's causing the problem. The line for the equalizer might be restricted. This would mean you would have to replace the equalizer. It is also possible that the valve is equipped with an improper equalizer. Under these conditions, you would have to replace the equalizer with one that was designed for your needs.

Gas in the line

You might get gas in your liquid line. This can happen due to a pressure drop in the line or an insufficient refrigerant charge. If this happens, check the refrigerant charge and add refrigerant if necessary. Make sure the size of the liquid line is what the manufacturer's specifications call for. You might have to clean the strainers and replace the filter drier. It is also a good idea to increase the head pressure or decrease the temperature to guarantee you have solid liquid refrigerant at the valve inlet. It might not be necessary to complete all of these steps, but one (or all) of them should solve your problem.

Clogged filter

A clogged filter screen can give you low suction pressure and high superheat. To solve this problem, all you have to do is clean the filter.

Superheat is too high

If the operating superheat is too high, you could experience low suction pressure and high superheat. If this is the case, simply adjust the superheat as you were told earlier.

Oil

If you use the wrong type of oil, your heat pump might suffer from the symptoms of low suction pressure and high superheat. To correct this, all you have to do is purge the system and install the proper type of oil.

Too small

It is possible that the orifice of the valve is too small. Replacing the valve with one that has a larger opening will correct this problem.

Plugged orifice

Sometimes wax and oil clogs the opening of the orifice. The wax and oil indicate that the wrong type of oil was used in the system. Purge the system and replace the oil content with a proper oil. You might also want to install a filter/drier to prevent moisture and dirt from blocking the orifice of the valve.

Power assembly

A faulty power assembly can inhibit the proper operation of your heat pump. When the power assembly goes on the blink, it is best to replace it.

The remote bulb

The gas-charged remote bulb of a valve might lose control due to the tubing for the bulb being colder than the bulb. The best solution to this problem is the replacement of the existing bulb with a "W" cross ambient power assembly.

Frost

If you have frost developing on your line, it is generally an indication of a restriction in the line. Frost or a temperature reading that is below normal indicates a good place to look for obstructions in the line.

Other probable causes

There are other probable causes of a low suction pressure and a high superheat. For example, an obstructed line could be the culprit. If the liquid line is too small, that could be the source of your problem. It is also conceivable that the suction line is too small. If the wrong type of oil was inducted into the system, it could be blocking the refrigerant flow. The solenoid valve might be too small or not operating properly. Valves that are too small or that are not opening fully could be at fault. There are, to be sure, many possible causes for a low suction pressure and a high superheat. Now, suppose you have low suction pressure and low superheat; how does that change your troubleshooting and repair procedures?

The compressor

If the compressor is oversized or running too fast, it could cause your system to have low suction pressure and low superheat. To correct this, you might have to install a different pulley on the compressor, one that would make it run at the proper speed. Another option might be to install a compressor capacity control.

Oil in the evaporator

An excessive build-up of oil in the evaporator could also create low suction pressure and low superheat. This is usually a result of the suction piping not being installed properly, but it might be necessary to install an oil separator. Check the piping to see that it will return oil properly. If it is not, change the piping arrangement so that it will. Should the piping already be installed properly, defer to the installation of an oil separator.

Ice

An ice build-up will generally mean that the evaporator is too small. To solve this problem, replace the evaporator with one that is of the proper size.

Uneven loading

Uneven loading of the evaporator, or inadequate loading, can be the reason why you are fighting with a low suction pressure and a low superheat. Inadequate distribution of air or brine flow is the cause in this case. To correct the problem, you must balance the evaporator's load distribution by providing a proper air or brine flow.

Poor distribution

Poor distribution in the evaporator will cause liquid to divert to the path of least resistance, so to speak. This can lead to throttling the valve prior to all passages receiving an adequate flow of refrigerant. When you suspect this as your problem, you should clamp a power assembly remote bulb to the free-draining suction line. The line must be cleaned before the bulb is attached. Then you should install a distributor that controls the flow. Take some time to balance the evaporator's load distribution, and your problem should be solved.

High suction

If you have a unit that is exhibiting a high suction pressure and a high superheat, there are only a few likely causes of the problem. The compressor might be too small. If it is, you must replace it with a larger compressor. The evaporator might be too large. This too requires a complete replacement, with an evaporator of the correct size. There is a chance that the compressor's discharge valves are leaking. In this case, you must repair or replace the valves. The fourth possibility is an unbalanced system. If the load is in excess of the design conditions, you must balance the load.

Low superheat

A low superheat with a high suction pressure can be caused by a number of things. The compressor might be too small, in which case, you must replace the compressor with one of the proper size. Your problem might be solved by simply adjusting the superheat setting. If the discharge valves associated with the compressor are leaking, they will have to be repaired or replaced. The expansion valve might be of the wrong size. This can lead to gas in the liquid line, requiring the replacement of the valve.

Low superheat

Let's say you have low superheat and high suction pressure. What is likely to cause this combination? The compressor could be too small. If it is, you must replace it with a larger unit. One of the first things to check out is the adjustment of the superheat. When this adjustment is too low, all you have to do is turn it up.

Compressor valves sometimes leak, as I've shown before. A leak in a compressor valve can cause low superheat and high suction pressure. Check the valves, and if they're leaking, repair or replace them. If your expansion valve is not sized properly, you might get gas in your liquid line. This, too, can cause a low superheat and a high suction pressure. To solve the problem, replace the valve with one that is of the proper size.

Moisture is a frequent enemy to heat pumps. If moisture freezes the expansion valve in an open position, low superheat and high suction pressure is likely. To solve this problem, you must thaw the valve. An easy way to do this involves wetting rags with hot water and applying them to the valve. Saturate the rags with hot water and continue wrapping the valve until the ice melts. Once you've solved the immediate problem, install a filter/drier to prevent the problem from reoccurring.

If the diaphragm of an automatic expansion valve breaks, it will result in a liquid feedback. This will call for replacement of the power assembly. Another cause of liquid feedback can be that the valve is sticking in an open position. You might have to replace the valve under these conditions, but it is also possible that a good cleaning will do the job. In any event, a filter should be installed to prevent future problems of this nature.

High discharge pressure

When a heat pump has a high discharge pressure, you might have too much refrigerant in the system. Check the refrigerant charge and purge it if necessary. If the condenser is dirty, you should clean it. Maybe the condenser or liquid receiver is too small. If this is the case, replace the equipment with a unit that is of the proper size.

You might find that air or noncondensables have entered the system. This will be cause to purge and recharge the entire system. A water valve that is out of adjustment can be causing your problem. If the cooling water is above the design temperature, you should increase the supply of water. It might even be necessary to install a larger valve.

Sometimes the cooling water is inadequate due to an inadequate supply of water or a bad water valve. If this is the case, start the pump and open the water valves. Should you find problems with the valves, repair or replace them.

The last likely cause of a high discharge pressure is the air-cooled condenser. It might not be getting enough air circulation. You should also check for worn belts and pulleys that might be slipping. If these conditions check out okay, inspect the blower motor to ensure that it is sized properly.

Erratic discharge pressure

An erratic discharge pressure with your heat pump could be related to one of six causes. The controls could be bad on the fluctuating discharge pressure. If they are, you will have to adjust, repair, or replace them. It might be necessary to replace the condensing water regulating valve, if it is bad. Another possibility is an improper charge of refrigerant. Check your refrigerant levels and add to it as needed.

If the cooling fan for the condenser cycle is out of whack, you must figure out why it is not operating properly and correct it. The cycling of the evaporative condenser is another consideration to take into account. Check the spray nozzles, coil surface, control circuits, and thermostat overloads. You might find that it will be necessary to clean the nozzles or coil surface. If any other defective equipment is detected, it should be replaced.

Erratic suction pressure

If you are experiencing an erratic suction pressure, check the superheat adjustment; it might need to be changed. If this isn't the cause of the problem, consider installing a P-trap on the suction line to provide a free-draining line. A restricted external equalizer line could also be causing your trouble. Check the line and replace it if necessary.

If the remote bulb location is not right, you could have erratic suction pressure. Check the location of the bulb, and change it if necessary. Remember to clean the suction line thoroughly before clamping the bulb into place at a new location.

If you are experiencing a change in pressure drop across the valve, you might have a faulty condensing water regulator. If so, replace it. A flood of backflow from the liquid refrigerant, caused by an uneven evaporator loading or an improper liquid distribution device is another possibility. Even an improperly mounted evaporator could be at fault. These problems could require you to remount the evaporator lines at a proper angle, replace the distributor, or to install the proper load distributor devices to balance the air velocity over the evaporator coils.

Inspect each valve in your system to ensure that they have their own equalizer line going directly to a proper location on the evaporator outlet. If they don't, rework the system so that they will.

As a last check, inspect spray nozzles, coil surface, control circuits, and thermostat overloads. The nozzles or coil surface might require cleaning. Control circuits and thermostat overloads might require replacement.

19

Know your limitations

Before you embark on the installation of a new heat pump, it is imperative that you know the limitations of your personal ability. Ego can get in the way of logic when it comes time to draw boundary lines for what a person can and can't do.

I encounter my own problems with limitations these days. When I was 18 years old and getting into the trades, I felt invincible. There didn't seem to be any physical task that I wouldn't tackle. Sore muscles were par for the course, but the pain went away in a few days and I kept on stroking. Nowadays, things are different. I've grown older, more experienced, smarter, and less able to do what I could do as a young buck. This is a fact of life that is hard for me to swallow.

I recently tore the rotator cuff in my right shoulder. When I went to see my doctor, I was told that this kind of injury in people under the age of thirty is usually a stretched muscle that will come back into shape after a short while and some use. However, for people over thirty, the same injury often results in a tear, as mine did, rather than a stretch, and the recovery time is much longer. Also, continuing to use the muscle at my age can result in some serious damage, rather than just working the kinks out. I've got to tell you, I'm not used to being old, I'm still in my thirties, and this type of injury is a real eye-opener to what's ahead.

Now my physical condition is not a factor in your heat pump installation, but your own limitations are. Just as I have to temper my old bulldog methods for beating through a job, you have to take stock of your qualifications as an HVAC installer. You must be honest with yourself; cheating will only serve to hurt you.

Installing a heat pump and putting it into operation is no minor task. The professionals that do this type of work train for a long time before they are competent to go out and do complete installations on their own. There is no reason to assume that you can read a book or two and be prepared for all of what you will be getting into when you tackle the installation of your own heat pump.

I can't tell you if you have the skills, determination, and ability to install your own heat pump. There will be some aspects of the job where you will have to take a try-and-see approach to your abilities. There are, however, probably some aspects of the work where you can make a sound judgment call on your own abilities. For example, if you don't know how to solder copper joints, you are at a definite disadvantage. A person who has never worked with electrical controls and wiring is going to have a tough time trying to complete a heat-pump installation. These are just examples, but there are several areas of an installation that might shoot up red flags to warn you of possible problems.

When you suspect that you are on the edge of your abilities, you have two choices. You can try to do the job and see what happens, or you can rely on professional help. The cost of professional help is not cheap, but it can be quite a bargain. The individual who wades into deep water without an ability to swim out of it is likely to need someone to come to the rescue. The same principle can be applied to you and your heat pump. If you venture into unknown waters, it is very possible that you will do more damage than good. This will not only require the expense of a professional to complete the task, but you might very well have to pay the pro to fix your mistakes. The cost of having someone repair work you've already done can be more expensive than paying them to do it in the first place.

The whole purpose of this chapter is to help you evaluate your skills in relation to a heat-pump installation. Some homeowners are capable of handling their jobs from start to finish, but most people need a little help here and there from professionals.

Designing

Designing your own heat-pump system is probably a bit much to take on by yourself. There are many aspects of this job that you can do, but some of the planning and design work is simply beyond the capabilities of most people. It should not, however, be a problem for you to determine what type of heat pump will best suit your needs. Most people will opt for an air-source unit, but with the information provided in this book, you have enough details to investigate alternative types of heat pumps.

Laying out the ductwork is a job that many homeowners can handle. Choosing a good location for the outside unit is not much of a bother. But, when it comes to sizing the system, professional help will probably be needed. You've seen how difficult it is to do an accurate load calculation and how important the proper sizing of the heat pump and ducts is. With so much riding on the job at hand, you will do well to consult professionals for this phase of the work.

The rough-in

The rough-in work required when installing a heat pump is not beyond the abilities of handy homeowners. If you are comfortable working with tools, you can probably get through the rough-in phase without any outside help. Some aspects of the duct work might require an extra set of hands, but the technical aspects of the job are not difficult to learn.

Setting the units

Setting the inside and outside units is a job that many homeowners can handle. You might need some help, (not necessarily professional help) getting the units in place, but once they are set in their permanent locations, you can do most of the work yourself.

Testing and starting

Testing and starting the new heat-pump system is an area where the assistance of a professional might be wise. Due to the work with refrigerants, electrical wiring, controls, gauges, and other devices that many people have never worked with before, the testing and starting of a new system can be a major responsibility. You do best by setting up the system yourself and then having a professional come in to look over your work, test the system, and put the system into operation.

Troubleshooting

Troubleshooting your heat pump is something you can probably do. If you have a basic knowledge of controls and electrical wiring, there is a good chance you can do your own troubleshooting and repairs. There are some situations where special equipment is required for troubleshooting and repairing heat-pump systems, and these areas of the work might require professional help. By reading through the troubleshooting and repair chapters it should be easy for you to decide what you can and can't do.

Don't venture beyond your limits

There are certainly some aspects of installing a heat pump that can lead to trouble. For instance, if you connect the wiring to the controls improperly, you could damage the controls. Then you not only have to pay a professional to wire the control correctly, you also have to buy a new control. There are many circumstances similar to the example I've just given you.

How can you keep out of trouble? You can do so by knowing your limitations and by not venturing in to work that you don't understand. When you read this book or a manufacturer's recommendations, you should have a complete understanding of what you are going to do before you do it. If you don't, trouble is likely to follow.

Assume that you have purchased precharged refrigerant tubing. You've read about installing the tubing, and you know the points on the equipment where the copper will be connected. However, you only skimmed the warnings about how easy it is to kink copper tubing. As you are feeding the tubing through the outside wall of your home, you grab one end of the copper and pull it. You didn't unroll the copper as you should have, and by pulling on one end while the rest of the tubing was coiled on the floor, you created a serious kink in the soft copper. Now what are you going to do? The only thing you can do is buy more tubing. If you leave the kink in the copper, you won't achieve a proper flow of refrigerant, and this will alter the operation of your heat pump. If you cut and splice the tubing you will lose your charge of refrigerant. This is a very simple example, but it is one that could happen quite easily.

If you are going to do the work yourself, you must understand how to do it. Even if you spend twice as much time learning how to do the job than what you spend actually performing the work, you will be better off. A lot of people skim instructions and hope for the best. If you take this approach, your negligence could prove to be an expensive lesson.

Calling for help

Calling for help from professionals is not always as easy as one might think. Sure, you can open up a local phone directory and call any number of professional contractors, but how will you know you are getting someone with the qualifications required for your job? Can you trust all the contractors who advertise in the phone book? Have you thought to ask if the contractor is licensed and insured to perform HVAC work? There is a lot more involved in finding a competent contractor than many people know.

I've worked in the trades for about 20 years. Much of my experience has been gained as a contractor. During these years, I've seen some pretty wild things happen in the field. Unfortunately, many of the memorable occasions have to do with homeowners who were abused by their contractors. Most contractors are reputable, but there are large numbers of contractors who are not interested in giving customers a fair shake. If you want quality work done for a reasonable price, you have to be willing to invest a little time in choosing the proper professionals.

An entire book could be written on how to choose the right contractor. Here, I can only give you a short course in the best ways to ensure that you get the quality service you want at fair prices.

Licenses

Most locations require heating and cooling professionals to have more than just a business license. Typically, such professionals must possess a trade license that indicates they are competent to perform the work they do. Just because someone has a license doesn't mean they are safe to hire, but it is a step in the right direction.

Call your local code enforcement office and see what the certification and licensing requirements are in your area. Once you know what is required, request to see proof that any contractor you are considering has met the qualifications necessary to be a legitimate professional in your area.

Insurance

Insurance is a big factor in choosing a contractor. You might think that anyone in the business would carry adequate insurance, but this just isn't true. A surprisingly large percentage of contractors fail to maintain adequate insurance coverage. I wouldn't even think about doing work without suitable insurance coverage, but there are a lot of contractors who feel insurance is a waste of money.

For your safety, you must make sure any contractor you engage is properly insured. At the least, this will mean that the contractor must have a good liability policy. If the contractor has employees, worker's compensation insurance can be an issue. Bonding is another consideration, however, it is not generally as important as a good liability policy.

When you are checking out contractors, don't believe everything they tell you. I hate to say it, but some contractors will lie to you about their insurance coverage. Demand that the contractor have the company who holds the insurance policy for the company send you a certificate of insurance. Never accept a photocopy given to you by the contractor. The certificate should come from the insurance company directly. Check the effective dates on the policy to make sure it is in force and will remain in force during the period of time needed to complete your job.

References

I have mixed feelings when it comes to references supplied by contractors. Obviously, a contractor is only going to give perspective customers the names of past customers who are satisfied. What about all the unsatisfied customers? Even with the glitch in the system, it is necessary to get references. The names you are given should be of customers who have had work similar to the type of work you want done performed for them by the contractor.

Don't just take a name list and look over it. Contact the customers and see if they were happy with the contractor. Do your best to make sure you weren't given names of people who are related to the contractor. Spend some time doing your homework.

Ask your friends

Ask your friends for recommendations pertaining to contractors who they have used. This type of reference base is much more reliable than the references given to you by some contractors. If your friends or neighbors have had good results with a specific contractor, it is likely you will too.

Watch out for bargains

Watch out for bargains when you are shopping prices with contractors. Contractors who are cheap are not necessarily a good deal. Many times you get what you pay for, so keep this in mind. If a price looks too good to be true, it probably is.

Permits

Most jurisdictions require a permit to be issued from the local code enforcement office prior to the installation of a new heating system.

These permits are generally made available to homeowner who will be doing their own work and to properly licensed professionals. If you get a contractor who is planning to do the job without a permit and the required code inspections, look out. You might save a little money in the beginning, but you might also be buying into a lot of grief.

Deposits

Many contractors will ask for a monetary deposit when an agreement for work is entered into. This is common practice, but it is not in your best interest. The contractors want the deposit because they want to operate with your money rather than theirs. They also want to make sure you are going to pay them. However, if you give a deposit, you are at risk. Suppose the contractor takes the money and disappears? It has happened before, many times. Avoid giving deposits of any substantial sum, and never give more than one-third of the contract price up front.

Why should you pay someone for work that hasn't been done? That is, in effect, what you are doing when you pay a deposit. I know it will be difficult to get a contractor to take on your work without a deposit, but keep the amount as low as possible, and avoid laying out cash before you have the goods or services provided.

In general

In general, you have to be careful who you hire. Just because a company has an impressive advertisement in the phone directory doesn't mean the company is stable, reliable, or even properly licensed. I can't emphasize enough how important it is to check out your potential contractors closely.

I wish we had the space to expand on the issue of unscrupulous contractors; there are many stories I could tell you. At least you now know that choosing a contractor is not as simple as calling the first phone number you run across in the phone directory.

Hopefully, this book has given you more knowledge about heat pumps than most people will ever seek to gain. While it is not the same as having a seasoned veteran working at your side, it will be of great help to you in working with your heat pump. Remember two things; don't hesitate to call a professional if you don't understand the job you're doing, and always follow the recommendations made by the manufacturer of your equipment. I wish you the very best of luck in your endeavors.

Index

Illustrations are in **boldface**.

About the Author

R. Dodge Woodson has had nearly 20 years experience as a homebuilder, contractor, master plumber, and real estate broker. He is also the author of *Home Plumbing Illustrated, Roofing Contractor: Start and Run a Money-Making Business*, and *The Master Plumber's Licensing Exam Guide*, among others. Woodson has also written articles for various magazines, including *Fur, Fish and Game* and *Outdoor Life*.

Other Bestsellers of Related Interest

HVAC Design Data Sourcebook
Robert O. Parmley
The complete, quick-and-easy sourcebook of fundamental HVAC information and data. Includes tabular data for calculating heating fuel used and up-to-date recommendations for energy conservation.
ISBN 0-07-048572-0 $49.50 Hard

HVAC Field Manual
Robert O. Parmely
A comprehensive, portable manual that covers all the essential technology on heating fuels and sources, heat loss, energy conservation, and more. Loaded with charts, tables, formulas, and conversions, this ready-reference book will help solve problems quickly and improve job performance.
ISBN 0-07-048524-0 $42.50 Hard

HVAC Mechanic: Start and Run a Money-Making Business
R. Dodge Woodson
a well-illustrated, helpful guide for those who want to turn their HVAC service knowledge into a profitable business. Provided here is practical advice on becoming your own boss, managing time and money, finding clients, establishing credit, and more.
ISBN 0-07-071776-1 $17.95 Paper

Heat and Thermodynamics
M.W. Zemansky and R. H. Dittman
Complete guide to heat and themodynamics theory and application. This book is the last word for technicians, students, and engineers alike on understanding heat properties and thermodynamic rules and how these apply to technical practice and design.
ISBN 0-07-072808-9 $62.50 Hard

How to Order

 Call 1-800-822-8158
24 hours a day,
7 days a week
in U.S. and Canada

 Mail this coupon to:
McGraw-Hill, Inc.
Blue Ridge Summit, PA
17294-0840

 Fax your order to:
717-794-5291

 EMAIL
70007.1531@COMPUSERVE.COM
COMPUSERVE: GO MH

Thank you for your order!

Shipping and Handling Charges

Order Amount	Within U.S.	Outside U.S.
Less than $15	$3.45	$5.25
$15.00 - $24.99	$3.95	$5.95
$25.00 - $49.99	$4.95	$6.95
$50.00 - and up	$5.95	$7.95

EASY ORDER FORM—
SATISFACTION GUARANTEED

Ship to:

Name _____

Address _____

City/State/Zip _____

Daytime Telephone No. _____

ITEM NO.	QUANTITY	AMT.

Method of Payment:

☐ Check or money order
   enclosed (payable to
   McGraw-Hill)

☐ [Cards] ☐ *VISA*

☐ MasterCard ☐ DISCOVER

Method of Payment:	Shipping & Handling charge from chart below
	Subtotal
	Please add applicable state & local sales tax
	TOTAL

Account No. ☐☐☐☐☐☐☐☐☐☐☐☐☐☐☐☐

Signature _____ Exp. Date ____
Order invalid without signature

**In a hurry? Call 1-800-822-8158 anytime,
day or night, or visit your local bookstore.**

Code = BC44ZNA